UNDERSTANDING PHYSICS

D0705428

PART 4

Karen Cummings
Rensselaer Polytechnic Institute
Southern Connecticut State University

Priscilla W. Laws
Dickinson College

Edward F. Redish
University of Maryland

Patrick J. Cooney
Millersville University

GUEST AUTHOR

Edwin F. Taylor
Massachusetts Institute of Technology

ADDITIONAL MEMBERS OF ACTIVITY BASED PHYSICS GROUP

David R. Sokoloff
University of Oregon

Ronald K. Thornton
Tufts University

Understanding Physics is based on *Fundamentals of Physics*
by David Halliday, Robert Resnick, and Jearl Walker.

WILEY

John Wiley & Sons, Inc.

This book is dedicated to Arnold Arons,
whose pioneering work in physics education
and reviews of early chapters have had
a profound influence on our work.

SENIOR ACQUISITIONS EDITOR   Stuart Johnson
SENIOR DEVELOPMENT EDITOR   Ellen Ford
MARKETING MANAGER   Bob Smith
SENIOR PRODUCTION EDITOR   Elizabeth Swain
SENIOR DESIGNER   Kevin Murphy
INTERIOR DESIGN   Circa 86, Inc.
COVER DESIGN   David Levy
COVER PHOTO   © Antonio M. Rosario/The Image Bank/Getty Images
ILLUSTRATION EDITOR   Anna Melhorn
PHOTO EDITOR   Hilary Newman

This book was set in 10/12 Times Ten Roman by Progressive and
printed and bound by Von Hoffmann Press. The cover was printed by Von Hoffmann Press.

This book is printed on acid free paper. ∞

Library of Congress Cataloging in Publication Data:

Understanding physics / Karen Cummings . . . [et al.]; with additional members of the
    Activity Based Physics Group.
        p. cm.
    Includes index.
    ISBN 0-471-46438-4 (pt. 4 : pbk. : acid-free paper)
    1. Physics.   I. Cummings, Karen.   II. Activity Based Physics Group.

QC23.2.U54 2004
530—dc21   2003053481

L.C. Call no.   Dewey Classification No.   L.C. Card No.

Printed in the United States of America

10 9 8 7 6 5 4 3 2 1

Preface

Welcome to *Understanding Physics*. This book is built on the foundations of the 6th Edition of Halliday, Resnick, and Walker's *Fundamentals of Physics* which we often refer to as HRW 6th. The HRW 6th text and its ancestors, first written by David Halliday and Robert Resnick, have been best-selling introductory physics texts for the past 40 years. It sets the standard against which many other texts are judged. You are probably thinking, "Why mess with success?" Let us try to explain.

Why a Revised Text?

A physics major recently remarked that after struggling through the first half of his junior level mechanics course, he felt that the course was now going much better. What had changed? Did he have a better background in the material they were covering now? "No," he responded. "I started reading the book before every class. That helps me a lot. I wish I had done it in Physics One and Two." Clearly, this student learned something very important. It is something most physics instructors wish they could teach all of their students as soon as possible. Namely, no matter how smart your students are, no matter how well your introductory courses are designed and taught, your students will master more physics if they learn how to read an "understandable" textbook carefully.

We know from surveys that the vast majority of introductory physics students do not read their textbooks carefully. We think there are two major reasons why: (1) many students complain that physics textbooks are impossible to understand and too abstract, and (2) students are extremely busy juggling their academic work, jobs, personal obligations, social lives and interests. So they develop strategies for passing physics without spending time on careful reading. We address both of these reasons by making our revision to the sixth edition of *Fundamentals of Physics* easier for students to understand and by providing the instructor with more **Reading Exercises** (formerly known as Checkpoints) and additional strategies for encouraging students to read the text carefully. Fortunately, we are attempting to improve a fine textbook whose active author, Jearl Walker, has worked diligently to make each new edition more engaging and understandable.

In the next few sections we provide a summary of how we are building upon HRW 6th and shaping it into this new textbook.

A Narrative That Supports Student Learning

One of our primary goals is to help students make sense of the physics they are learning. We cannot achieve this goal if students see physics as a set of disconnected mathematical equations that each apply only to a small number of specific situations. We stress conceptual and qualitative understanding and continually make connections between mathematical equations and conceptual ideas. We also try to build on ideas that students can be expected to already understand, based on the resources they bring from everyday experiences.

In *Understanding Physics* we have tried to tell a story that flows from one chapter to the next. Each chapter begins with an introductory section that discusses why new topics introduced in the chapter are important, explains how the chapter builds on previous chapters, and prepares students for those that follow. We place explicit emphasis on basic concepts that recur throughout the book. We use extensive forward and backward referencing to reinforce connections between topics. For example, in the introduction of Chapter 16 on Oscillations we state: "Although your study of simple harmonic motion will enhance your understanding of mechanical systems it is also vital to understanding the topics in electricity and magnetism encountered in Chapters 30-37. Finally, a knowledge of SHM provides a basis for understanding the wave nature of light and how atoms and nuclei absorb and emit energy."

Emphasis on Observation and Experimentation

Observations and concrete everyday experiences are the starting points for development of mathematical expressions. Experiment-based theory building is a major feature of the book. We build ideas on experience that students either already have or can easily gain through careful observation.

Whenever possible, the physical concepts and theories developed in *Understanding Physics* grow out of simple observations or experimental data that can be obtained in typical introductory physics laboratories. We want our readers to develop the habit of asking themselves: What do our observations, experiences and data imply about the natural laws of physics? How do we know a given statement is true? Why do we believe we have developed correct models for the world?

Toward this end, the text often starts a chapter by describing everyday observations with which students are familiar. This makes *Understanding Physics* a text that is both relevant to students' everyday lives and draws on existing student knowledge. We try to follow Arnold Arons' principle "idea first, name after." That is, we make every attempt to begin a discussion by using everyday language to describe common experiences. Only then do we introduce formal physics terminology to represent the concepts being discussed. For example, everyday pushes, pulls, and their impact on the motion of an object are discussed before introducing the term "force" or Newton's Second Law. We discuss how a balloon shrivels when placed in a cold environment and how a pail of water cools to room temperature before introducing the ideal gas law or the concept of thermal energy transfer.

The "idea first, name after" philosophy helps build patterns of association between concepts students are trying to learn and knowledge they already have. It also helps students reinterpret their experiences in a way that is consistent with physical laws.

Examples and illustrations in *Understanding Physics* often present data from modern computer-based laboratory tools. These tools include computer-assisted data acquisition systems and digital video analysis software. We introduce students to these tools at the end of Chapter 1. Examples of these techniques are shown in Figs. P-1 and P-2 (on the left) and Fig. P-3 on the next page. Since many instructors use these computer tools in the laboratory or in lecture demonstrations, these tools are part of the introductory physics experience for more and more of our students. The use of real data has a number of advantages. It connects the text to the students' experience in other parts of the course and it connects the text directly to real world experience. Regardless of whether data acquisition and analysis tools are used in the student's own laboratory, our use of realistic rather that idealized data helps students develop an appreciation of the role that data evaluation and analysis plays in supporting theory.

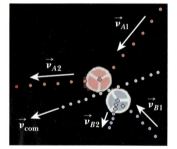

FIGURE P-1 ■ A video analysis shows that the center of mass of a two-puck system moves at a constant velocity.

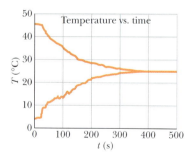

FIGURE P-2 ■ Electronic temperature sensors reveal that if equal amounts of hot and cold water mix the final temperature is the average of the initial temperatures.

FIGURE P-3 ■ A video analysis of human motion reveals that in free fall the center of mass of an extended body moves in a parabolic path under the influence of the Earth's gravitational force.

Using Physics Education Research

In re-writing the text we have taken advantage of two valuable findings of physics education research. One is the identification of concepts that are especially difficult for many students to learn. The other is the identification of active learning strategies to help students develop a more robust understanding of physics.

Addressing Learning Difficulties

Extensive scholarly research exists on the difficulties students have in learning physics.[1] We have made a concerted effort to address these difficulties. In *Understanding Physics,* issues that are known to confuse students are discussed with care. This is true even for topics like the nature of force and its effect on velocity and velocity changes that may seem trivial to professional physicists. We write about subtle, often counter-intuitive topics with carefully chosen language and examples designed to draw out and remediate common alternative student conceptions. For example, we know that students have trouble understanding passive forces such as normal and friction forces.[2] How can a rigid table exert a force on a book that rests on it? In Section 6-4 we present an idealized model of a solid that is analogous to an inner spring mattress with the repulsion forces between atoms acting as the springs. In addition, we invite our readers to push on a table with a finger and experience the fact that as they push harder on the table the table pushes harder on them in the opposite direction.

FIGURE P-4 ■ Compressing an innerspring mattress with a force. The mattress exerts an oppositely directed force, with the same magnitude, back on the finger.

Incorporating Active Learning Opportunities

We designed *Understanding Physics* to be more interactive and to foster thoughtful reading. We have retained a number of the excellent Checkpoint questions found at the end of HRW 6th chapter sections. We now call these questions **Reading Exercises.** We have created many new Reading Exercises that require students to reflect on the material in important chapter sections. For example, just after reading Section 6-2 that introduces the two-dimensional free-body diagram, students encounter Reading Exercise 6-1. This multiple-choice exercise requires students to identify the free-body diagram for a helicopter that experiences three non-collinear forces. The distractors were based on common problems students have with the construction of free-body diagrams. When used in "Just-In-Time Teaching" assignments or for in-class group discussion, this type of reading exercise can help students learn a vital problem solving skill as they read.

---

[1] L. C. McDermott and E. F. Redish, "Resource Letter PER-1: Physics Education Research," *Am. J. Phys.* **67**, 755-767 (1999)

[2] John J. Clement, "Expert novice similarities and instruction using analogies," *Int. J. Sci. Ed. 20,* 1271-1286 (1998)

We also created a set of **Touchstone Examples.** These are carefully chosen sample problems that illustrate key problem solving skills and help students learn how to use physical reasoning and concepts as an essential part of problem solving. We selected some of these touchstone examples from the outstanding collection of sample problems in HRW 6th and we created some new ones. In order to retain the flow of the narrative portions of each chapter, we have reduced the overall number of sample problems to those necessary to exemplify the application of fundamental principles. Also, we chose touchstone examples that require students to combine conceptual reasoning with mathematical problem-solving skills. Few, if any, of our touchstone examples are solvable using simple "plug-and-chug" or algorithmic pattern matching techniques.

Alternative problems have been added to the extensive, classroom tested end-of-chapter problem sets selected from HRW 6th. The design of these new problems are based on the authors' knowledge of research on student learning difficulties. Many of these new problems require careful qualitative reasoning. They explicitly connect conceptual understanding to quantitative problem solving. In addition, estimation problems, video analysis problems, and "real life" or "context rich" problems have been included.

The organization and style of *Understanding Physics* has been modified so that it can be easily used with other research-based curricular materials that make up what we call *The Physics Suite*. The *Suite* and its contents are explained at length at the end of this preface.

Reorganizing for Coherence and Clarity

For the most part we have retained the organization scheme inherited from HRW 6th. Instructors are familiar with the general organization of topics in a typical course sequence in calculus-based introductory physics texts. In fact, ordering of topics and their division into chapters is the same for 27 of the 38 chapters. The order of some topics has been modified to be more pedagogically coherent. Most of the reorganization was done in Chapters 3 through 10 where we adopted a sequence known as *New Mechanics*. In addition, we decided to move HRW 6th Chapter 25 on capacitors so it becomes the last chapter on DC circuits. Capacitors are now introduced in Chapter 28 in *Understanding Physics*.

The New Mechanics Sequence

HRW 6th and most other introductory textbooks use a familiar sequence in the treatment of classical mechanics. It starts with the development of the kinematic equations to describe constantly accelerated motion. Then two-dimensional vectors and the kinematics of projectile motion are treated. This is followed by the treatment of dynamics in which Newton's Laws are presented and used to help students understand both one- and two-dimensional motions. Finally energy, momentum conservation, and rotational motion are treated.

About 12 years ago when Priscilla Laws, Ron Thornton, and David Sokoloff were collaborating on the development of research-based curricular materials, they became concerned about the difficulties students had working with two-dimensional vectors and understanding projectile motion before studying dynamics.

At the same time Arnold Arons was advocating the introduction of the concept of momentum before energy.[3] Arons argued that (1) the momentum concept is simpler than the energy concept, in both historical and modern contexts and (2) the study

---

[3] Private Communication between Arnold Arons and Priscilla Laws by means of a document entitled "Preliminary Notes and Suggestions," August 19, 1990; and Arnold Arons, *Development of Concepts of Physics* (Addison-Wesley, Reading MA, 1965)

of momentum conservation entails development of the concept of center-of-mass which is needed for a proper development of energy concepts. Additionally, the impulse-momentum relationship is clearly an alternative statement of Newton's Second Law. Hence, its placement immediately after the coverage of Newton's laws is most natural.

In order to address these concerns about the traditional mechanics sequence, a small group of physics education researchers and curriculum developers convened in 1992 to discuss the introduction of a new order for mechanics.[4] One result of the conference was that Laws, Sokoloff, and Thornton have successfully incorporated a new sequence of topics in the mechanics portions of various curricular materials that are part of the Physics Suite discussed below.[5] These materials include *Workshop Physics*, the *RealTime Physics Laboratory Module in Mechanics*, and the *Interactive Lecture Demonstrations*. This sequence is incorporated in this book and has required a significant reorganization and revisions of HRW 6th Chapters 2 through 10.

The New Mechanics sequence incorporated into Chapters 2 through 10 of understanding physics includes:

- Chapter 2: One-dimensional kinematics using constant horizontal accelerations and vertical free fall as applications.

- Chapter 3: The study of one-dimensional dynamics begins with the application of Newton's laws of motion to systems with one or more forces acting along a single line. Readers consider observations that lead to the postulation of "gravity" as a constant invisible force acting vertically downward.

- Chapter 4: Two-dimensional vectors, vector displacements, unit vectors and the decomposition of vectors into components are treated.

- Chapter 5: The study of kinematics and dynamics is extended to two-dimensional motions with forces along only a single line. Examples include projectile motion and circular motion.

- Chapter 6: The study of kinematics and dynamics is extended to two-dimensional motions with two-dimensional forces.

- Chapters 7 & 8: Topics in these chapters deal with impulse and momentum change, momentum conservation, particle systems, center of mass, and the motion of the center-of-mass of an isolated system.

- Chapters 9 & 10: These chapters introduce kinetic energy, work, potential energy, and energy conservation.

Just-in-Time Mathematics

In general, we introduce mathematical topics in a "just-in-time" fashion. For example, we treat one-dimensional vector concepts in Chapter 2 along with the development of one-dimensional velocity and acceleration concepts. We hold the introduction of two- and three-dimensional vectors, vector addition and decomposition until Chapter 4, immediately before students are introduced to two-dimensional motion and forces in Chapters 5 and 6. We do not present vector products until they are needed. We wait to introduce the dot product until Chapter 9 when the concept of physical work is presented. Similarly, the cross product is first presented in Chapter 11 in association with the treatment of torque.

[4] The New Mechanics Conference was held August 6-7, 1992 at Tufts University. It was attended by Pat Cooney, Dewey Dykstra, David Hammer, David Hestenes, Priscilla Laws, Suzanne Lea, Lillian McDermott, Robert Morse, Hans Pfister, Edward F. Redish, David Sokoloff, and Ronald Thornton.

[5] Laws, P. W. "A New Order for Mechanics" pp. 125-136, *Proceedings of the Conference on the Introductory Physics Course*, Rensselaer Polytechnic Institute, Troy New York, May 20-23, Jack Wilson, Ed. 1993 (John Wiley & Sons, New York 1997)

Notation Changes

Mathematical notation is often confusing, and ambiguity in the meaning of a mathematical symbol can prevent a student from understanding an important relationship. It is also difficult to solve problems when the symbols used to represent different quantities are not distinctive. Some key features of the new notation include:

- We adhere to recent notation guidelines set by the U.S. National Institute of Standard and Technology Special Publication 811 (SP 811).

- We try to balance our desire to use familiar notation and our desire to avoid using the same symbol for different variables. For example, p is often used to denote momentum, pressure, and power. We have chosen to use lower case p for momentum and capital P for pressure since both variables appear in the kinetic theory derivation. But we stick with the convention of using capital P for power since it does not commonly appear side by side with pressure in equations.

- We denote vectors with an arrow instead of bolding so handwritten equations can be made to look like the printed equations.

- We label each vector component with a subscript that explicitly relates it to its coordinate axis. This eliminates the common ambiguity about whether a quantity represents a magnitude which is a scalar or a vector component which is not a scalar.

- We often use subscripts to spell out the names of objects that are associated with mathematical variables even though instructors and students will tend to use abbreviations. We also stress the fact that one object is exerting a force on another with an arrow in the subscript. For example, the force exerted by a rope on a block would be denoted as $\vec{F}_{\text{rope}\rightarrow\text{block}}$.

Our notation scheme is summarized in more detail in Appendix A4.

Encouraging Text Reading

We have described a number of changes that we feel will improve this textbook and its readability. But even the best textbook in the world is of no help to students who do not read it. So it is important that instructors make an effort to encourage busy students to develop effective reading habits. In our view the single most effective way to get students to read this textbook is to assign appropriate reading, reading exercises, and other reading questions after every class. Some effective ways to follow up on reading question assignments include:

1. Employ a method called "Just-In-Time-Teaching" (or JiTT) in which students submit their answers to questions about reading before class using just plain email or one of the many available computer based homework systems (Web Assign or E-Grade for example). You can often read enough answers before class to identify the difficult questions that need more discussion in class;

2. Ask students to bring the assigned questions to class and use the answers as a basis for small group discussions during the class period;

3. Assign multiple choice questions related to each section or chapter that can be graded automatically with a computer-based homework system; and

4. Require students to submit chapter summaries. Because this is a very effective assignment, we intentionally avoided doing chapter summaries for students.

Obviously, all of these approaches are more effective when students are given some credit for doing them. Thus you should arrange to grade all, or a random sample, of the submissions as incentives for students to read the text and think about the answers to Reading Exercises on a regular basis.

The Physics Suite

In 1997 and 1998, Wiley's physics editor, Stuart Johnson, and an informally constituted group of curriculum developers and educational reformers known as the *Activity Based Physics Group* began discussing the feasibility of integrating a broad array of curricular materials that are physics education research-based. This led to the assembly of an *Activity Based Physics Suite* that includes this textbook. The *Physics Suite* also includes materials that can be combined in different ways to meet the needs of instructors working in vastly different learning environments. The *Interactive Lecture Demonstration Series*[6] is designed primarily for use in lecture sessions. Other *Suite* materials can be used in laboratory settings including the *Workshop Physics Activity Guide,*[7] the *Real Time Physics Laboratory* modules,[8] and *Physics by Inquiry.*[9] Additional elements in the collection are suitable for use in recitation sessions such as the University of Washington *Tutorials in Introductory Physics* (available from Prentice Hall)[10] and a set of *Quantitative Tutorials*[11] developed at the University of Maryland. The *Activity Based Physics Suite* is rounded out with a collection of thinking problems developed at the University of Maryland. In addition to this **Understanding Physics** text, the Physics Suite elements include:

1. **Teaching Physics with the Physics Suite** by Edward F. Redish (University of Maryland). This book is not only the "Instructors Manual" for *Understanding Physics*, but it is also a book for anyone who is interested in learning about recent developments in physics education. It is a handbook with a variety of tools for improving both teaching and learning of physics—from new kinds of homework and exam problems, to surveys for figuring out what has happened in your class, to tools for taking and analyzing data using computers and video. The book comes with a Resource CD containing 14 conceptual and 3 attitude surveys, and more than 250 thinking problems covering all areas of introductory physics, resource materials from commercial vendors on the use of computerized data acquisition and video, and a variety of other useful reference materials. (Instructors can obtain a complimentary copy of the book and Resource CD, from John Wiley & Sons.)

2. **RealTime Physics** by David Sokoloff (University of Oregon), Priscilla Laws (Dickinson College), and Ronald Thornton (Tufts University). *RealTime Physics* is a set of laboratory materials that uses computer-assisted data acquisition to help students build concepts, learn representation translation, and develop an understanding of the empirical base of physics knowledge. There are three modules in the collection: Module 1: Mechanics (12 labs), Module 2: Heat and Thermodynamics (6 labs), and Module 3: Electric Circuits (8 labs). (Available both in print and in electronic form on *The Physics Suite CD.*)

---

[6]David R. Sokoloff and Ronald K. Thornton, "Using Interactive Lecture Demonstrations to Create an Active Learning Environment." *The Physics Teacher*, **35**, 340-347, September 1997.

[7]Priscilla W. Laws, *Workshop Physics Activity Guide*, Modules 1-4 w/ Appendices (John Wiley & Sons, New York, 1997).

[8]David R. Sokoloff, *RealTime Physics*, Modules 1-2, (John Wiley & Sons, New York, 1999).

[9]Lillian C. McDermott and the Physics Education Group at the University of Washington, *Physics by Inquiry* (John Wiley & Sons, New York, 1996).

[10]Lillian C. McDermott, Peter S. Shaffer, and the Physics Education Group at the University of Washington, *Tutorials in Introductory Physics*, First Edition (Prentice-Hall, Upper Saddle River, NJ, 2002).

[11]Richard N. Steinberg, Michael C. Wittmann, and Edward F. Redish, "Mathematical Tutorials in Introductory Physics," in, *The Changing Role Of Physics Departments In Modern Universities*, Edward F. Redish and John S. Rigden, editors, AIP Conference Proceedings **399**, (AIP, Woodbury NY, 1997), 1075-1092.

3. **Interactive Lecture Demonstrations** by David Sokoloff (University of Oregon) and Ronald Thornton (Tufts University). ILDs are worksheet-based guided demonstrations designed to focus on fundamental principles and address specific naïve conceptions. The demonstrations use computer-assisted data acquisition tools to collect and display high quality data in real time. Each ILD sequence is designed for delivery in a single lecture period. The demonstrations help students build concepts through a series of instructor led steps involving prediction, discussions with peers, viewing the demonstration and reflecting on its outcome. The ILD collection includes sequences in mechanics, thermodynamics, electricity, optics and more. (Available both in print and in electronic form on *The Physics Suite CD.*)

4. **Workshop Physics** by Priscilla Laws (Dickinson College). *Workshop Physics* consists of a four part activity guide designed for use in calculus-based introductory physics courses. Workshop Physics courses are designed to replace traditional lecture and laboratory sessions. Students use computer tools for data acquisition, visualization, analysis and modeling. The tools include computer-assisted data acquisition software and hardware, digital video capture and analysis software, and spreadsheet software for analytic mathematical modeling. Modules include classical mechanics (2 modules), thermodynamics & nuclear physics, and electricity & magnetism. (Available both in print and in electronic form on *The Physics Suite CD.*)

5. **Tutorials in Introductory Physics** by Lillian C. McDermott, Peter S. Shaffer and the Physics Education Group at the University of Washington. These tutorials consist of a set of worksheets designed to supplement instruction by lectures and textbook in standard introductory physics courses. Each tutorial is designed for use in a one-hour class session in a space where students can work in small groups using simple inexpensive apparatus. The emphasis in the tutorials is on helping students deepen their understanding of critical concepts and develop scientific reasoning skills. There are tutorials on mechanics, electricity and magnetism, waves, optics, and other selected topics. (Available in print from Prentice Hall, Upper Saddle River, New Jersey.)

6. **Physics by Inquiry** by Lillian C. McDermott and the Physics Education Group at the University of Washington. This self-contained curriculum consists of a set of laboratory-based modules that emphasize the development of fundamental concepts and scientific reasoning skills. Beginning with their observations, students construct a coherent conceptual framework through guided inquiry. Only simple inexpensive apparatus and supplies are required. Developed primarily for the preparation of precollege teachers, the modules have also proven effective in courses for liberal arts students and for underprepared students. The amount of material is sufficient for two years of academic study. (Available in print.)

7. **The Activity Based Physics Tutorials** by Edward F. Redish and the University of Maryland Physics Education Research Group. These tutorials, like those developed at the University of Washington, consist of a set of worksheets developed to supplement lectures and textbook work in standard introductory physics courses. But these tutorials integrate the computer software and hardware tools used in other Suite elements including computer data acquisition, digital video analysis, simulations, and spreadsheet analysis. Although these tutorials include a range of classical physics topics, they also include additional topics in modern physics. (Available only in electronic form on *The Physics Suite CD.*)

8. **The Understanding Physics Video CD for Students** by Priscilla Laws, et. al.: This CD contains a collection of the video clips that are introduced in *Understanding Physics* narrative and alternative problems. The CD includes a number of Quick-Time movie segments of physical phenomena along with the QuickTime player

software. Students can view video clips as they read the text. If they have video analysis software available, they can reproduce data presented in text graphs or complete video analyses based on assignments designed by instructors.

9. **WPTools** by Priscilla Laws and Patrick Cooney: These tools consist of a set of macros that can be loaded with Microsoft Excel software that allow students to graph data transferred from computer data acquisition software and video analysis software more easily. Students can also use the *WPTools* to analyze numerical data and develop analytic mathematical models.

10. **The Physics Suite CD.** This CD contains a variety of the Suite Elements in electronic format (Microsoft Word files). The electronic format allows instructors to modify and reprint materials to better fit into their individual course syllabi. The CD contains much useful material including complete electronic versions of the following: *RealTime Physics, Interactive Lecture Demonstrations, Workshop Physics, Activity Based Physics Tutorials.*

A Final Word to the Instructor

Over the past decade we have learned how valuable it is for us as teachers to focus on what most students actually need to do to learn physics, and how valuable it can be for students to work with research-based materials that promote active learning. We hope you and your students find this book and some of the other *Physics Suite* materials helpful in your quest to make physics both more exciting and understandable to your students.

Supplements for Use with Understanding Physics

Instructor Supplements

1. **Instructor's Solution Manual** prepared by Anand Batra (Howard University). This manual provides worked-out solutions for most of the end-of-chapter problems.

2. **Test Bank** by J. Richard Christman (U. S. Coast Guard Academy). This manual includes more than 2500 multiple-choice questions adapted from HRW 6th. These items are also available in the *Computerized Test Bank* (see below).

3. **Instructor's Resource CD.** This CD contains:
   - The entire *Instructor's Solutions Manual* in both Microsoft Word© (IBM and Macintosh) and PDF files.
   - A *Computerized Test Bank,* for use with both PCs and Macintosh computers with full editing features to help you customize tests.
   - All text illustrations, suitable for classroom projection, printing, and web posting.

4. **Online Homework and Quizzing:** *Understanding Physics* supports WebAssign and eGrade, two programs that give instructors the ability to deliver and grade homework and quizzes over the Internet.

Student Supplements

1. **Student Study Guide** by J. Richard Christman (U. S. Coast Guard Academy). This student study guide provides chapter overviews, hints for solving selected end-of-chapter problems, and self-quizzes.

2. **Student Solutions Manual** by J. Richard Christman (U. S. Coast Guard Academy). This manual provides students with complete worked-out solutions for approximately 450 of the odd-numbered end-of-chapter problems.

Acknowledgements

Many individuals helped us create this book. The authors are grateful to the individuals who attended the weekend retreats at Airlie Center in 1997 and 1998 and to our editor, Stuart Johnson and to John Wiley & Sons for sponsoring the sessions. It was in these retreats that the ideas for *Understanding Physics* crystallized. We are grateful to Jearl Walker, David Halliday and Bob Resnick for graciously allowing us to attempt to make their already fine textbook better.

The authors owe special thanks to Sara Settlemyer who served as an informal project manager for the past few years. Her contributions included physics advice (based on her having completed Workshop Physics courses at Dickinson College), her use of Microsoft Word, Adobe Illustrator, Adobe Photoshop and Quark XPress to create the manuscript and visuals for this edition, and skillful attempts to keep our team on task—a job that has been rather like herding cats.

Karen Cummings: I would like to say "Thanks!" to: Bill Lanford (for endless advice, use of the kitchen table and convincing me that I really could keep the same address for more than a few years in a row), Ralph Kartel Jr. and Avery Murphy (for giving me an answer when people asked why I was working on a textbook), Susan and Lynda Cummings (for the comfort, love and support that only sisters can provide), Jeff Marx, Tim French and the poker crew (for their friendship and laughter), my colleagues at Southern Connecticut and Rensselaer, especially Leo Schowalter, Jim Napolitano and Jack Wilson (for the positive influence you have had on my professional life) and my students at Southern Connecticut and Rensselaer, Ron Thornton, Priscilla Laws, David Sokoloff, Pat Cooney, Joe Redish, Ken and Pat Heller and Lillian C. McDermott (for helping me learn how to teach).

Priscilla Laws: First of all I would like thank my husband and colleague Ken Laws for his quirky physical insights, for the Chapter 11 Kneecap puzzler, for the influence of his physics of dance work on this book, and for waiting for me countless times while I tried to finish "just one more thing" on this book. Thanks to my daughter Virginia Jackson and grandson Adam for all the fun times that keep me sane. My son Kevin Laws deserves special mention for sharing his creativity with us—best exemplified by his murder mystery problem, *A(dam)nable Man*, reprinted here as problem 5-68. I would like to thank Juliet Brosing of Pacific University who adapted many of the Workshop Physics problems developed at Dickinson for incorporation into the alternative problem collection in this book. Finally, I am grateful to my Dickinson College colleagues Robert Boyle, Kerry Browne, David Jackson, and Hans Pfister for advice they have given me on a number of topics.

Joe Redish: I would like to thank Ted Jacobsen for discussions of our chapter on relativity and Dan Lathrop for advice on the sources of the Earth's magnetic field, as well as many other of my colleagues at the University of Maryland for discussions on the teaching of introductory physics over many years.

Pat Cooney: I especially thank my wife Margaret for her patient support and constant encouragement and I am grateful to my colleagues at Millersville University: John Dooley, Bill Price, Mike Nolan, Joe Grosh, Tariq Gilani, Conrad Miziumski, Zenaida Uy, Ned Dixon, and Shawn Reinfried for many illuminating conversations.

We also appreciate the absolutely essential role many reviewers and classroom testers played. We took our reviewers very seriously. Several reviewers and testers deserve special mention. First and foremost is Arnold Arons who managed to review 29 of the 38 chapters either from the original HRW 6th material or from our early drafts before he passed away in February 2001. Vern Lindberg from the Rochester

Institute of Technology deserves special mention for his extensive and very insightful reviews of most of our first 18 chapters. Ed Adelson from Ohio State did a particularly good job reviewing most of our electricity chapters. Classroom tester Maxine Willis from Gettysburg Area High School deserves special recognition for compiling valuable comments that her advanced placement physics students made while class testing Chapters 1-12 of the preliminary version. Many other reviewers and class testers gave us useful comments in selected chapters.

Class Testers

Gary Adams
Rensselaer Polytechnic Institute

Marty Baumberger
Chestnut Hill Academy

Gary Bedrosian
Rensselaer Polytechnic Institute

Joseph Bellina,
Saint Mary's College

Juliet W. Brosing
Pacific University

Shao-Hsuan Chiu
Frostburg State

Chad Davies
Gordon College

Hang Deng-Luzader
Frostburg State

John Dooley
Millersville University

Diane Dutkevitch
Yavapai College

Timothy Hayes
Rensselaer Polytechnic Institute

Brant Hinrichs
Drury College

Kurt Hoffman
Whitman College

James Holliday
John Brown University

Michael Huster
Simpson College

Dennis Kuhl
Marietta College

John Lindberg
Seattle Pacific University

Vern Lindberg
Rochester Institute of Technology

Stephen Luzader
Frostburg State

Dawn Meredith
University of New Hampshire

Larry Robinson
Austin College

Michael Roth
University of Northern Iowa

John Schroeder
Rensselaer Polytechnic Institute

Cindy Schwarz
Vassar College

William Smith
Boise State University

Dan Sperber
Rensselaer Polytechnic Institute

Roger Stockbauer
Louisiana State University

Paul Stoler
Rensselaer Polytechnic Institute

Daniel F. Styer
Oberlin College

Rebecca Surman
Union College

Robert Teese
Muskingum College

Maxine Willis
Gettysburg Area High School

Gail Wyant
Cecil Community College

Anne Young
Rochester Institute of Technology

David Ziegler
Sedro-Woolley High School

Reviewers

Edward Adelson
Ohio State University

Arnold Arons
University of Washington

Arun Bansil
Northeastern University

Chadan Djalali
University of South Carolina

William Dawicke
Milwaukee School of Engineering

Robert Good
California State University-Hayware

Harold Hart
Western Illinois University

Harold Hastings
Hofstra University

Laurent Hodges
Iowa State University

Robert Hilborn
Amherst College

Theodore Jacobson
University of Maryland

Leonard Kahn
University of Rhode Island

Stephen Kanim
New Mexico State University

Hamed Kastro
Georgetown University

Debora Katz
U. S. Naval Academy

Todd Lief
Cloud Community College

Vern Lindberg
Rochester Institute of Technology

Mike Loverude
California State University-Fullerton

Robert Luke
Boise State University

Robert Marchini
Memphis State University

Tamar More
Portland State University

Gregor Novak
U. S. Air Force Academy

Jacques Richard
Chicago State University

Cindy Schwarz
Vassar College

Roger Sipson
Moorhead State University

George Spagna
Randolf-Macon College

Gay Stewart
University of Arkansas-Fayetteville

Sudha Swaminathan
Boise State University

We would like to thank our proof readers Georgia Mederer and Ernestine Franco, our copyeditor Helen Walden, and our illustrator Julie Horan.

Last but not least we would like to acknowledge the efforts of the Wiley staff; Senior Acquisitions Editor, Stuart Johnson, Ellen Ford (Senior Development Editor), Justin Bow (Program Assistant), Geraldine Osnato (Project Editor), Elizabeth Swain (Senior Production Editor), Hilary Newman (Senior Photo Editor), Anna Melhorn (Illustration Editor), Kevin Murphy (Senior Designer), and Bob Smith (Marketing Manager). Their dedication and attention to endless details was essential to the production of this book.

Brief Contents

Contents

34 | Electromagnetic Waves

A comet is often referred to as a dirty snowball. As a comet swings around the Sun, ice on its surface vaporizes, releasing trapped dust and charged particles. The "solar wind," mostly consisting of protons streaming away from the Sun, forces the charged particles released by the comet into a straight "tail" that points radially away from the Sun. The dust continues to travel in the comet's orbit.

Why does most of the dust released by the comet remain in the tail of dust on the right in the photograph?

The answer is in this chapter.

FIGURE 34-1 ▪ James Clerk Maxwell (1831–1879).

34-1 Introduction

At the end of Chapter 31 we presented James Clerk Maxwell's (Fig. 34-1) four equations that synthesized 50 years of research on electricity and magnetism. Maxwell's equations are compact and elegant in the way that they reveal the inherent symmetry between electrical and magnetic phenomena. But their importance also stems from the fact that they led Maxwell and others to predict a wide range of new phenomena. The design of the communication systems that support radio and television broadcasting, cellular phones, and the Internet are all informed by Maxwell's equations. So is much of our understanding of the nature of light. For these reasons, Maxwell's equations are considered to be the crowning achievement of 19th-century theoretical physics.

One of the most incredible outcomes of Maxwell's work was his prediction of electromagnetic waves in 1864, long before investigators were able to generate and detect them. In fact, Maxwell's hypothesis was not taken seriously until almost 25 years later, when Heinrich Hertz first generated and detected electromagnetic waves. We begin this chapter by reviewing the remarkable chain of reasoning that led Maxwell to postulate the existence of this yet unknown type of wave. We also consider why it took Hertz until 1887 to confirm Maxwell's prediction. Next we describe how the electromagnetic wave pulses and continuous waves used for radio transmission are generated. We also examine how and why the orientation of radio and TV antennas is related to an idea called "polarization" of electromagnetic waves.

Maxwell predicted that all electromagnetic waves would move at a speed that was quite close to the measured value for the speed of visible light in air. For this reason he correctly asserted from the beginning that visible light was an electromagnetic wave. We can use what we learn about electromagnetic waves in this chapter to build a foundation for our study of optics in Chapter 35, where we will learn about how light waves interact with lenses and mirrors to form images.

34-2 Maxwell's Prediction of Electromagnetism

Maxwell's prediction of the electromagnetic wave was a result of the way in which he rewrote and reinterpreted the mathematical expressions for experimentally determined laws named after Faraday and Ampère. We start with a review of these two laws.

Generating Fields in the Absence of Conductors

Faraday's law

$$\mathcal{E} = -\frac{d\Phi^{\text{mag}}}{dt} \qquad \text{(Eq. 31-6)}$$

was originally based on his measurements of the electric currents induced *within a conducting loop* when the magnetic flux enclosed by it changes. The fact that an electric field is needed to produce the current in the conductor led Maxwell to a bold conjecture. He predicted that an electric field would *always* be induced in the region of varying magnetic flux regardless of whether or not a conducting loop was present. Thus, Faraday's law in the more general form shown below should probably be called the Faraday–Maxwell law:

$$\mathcal{E} = \oint \vec{E} \cdot d\vec{s} = -\frac{d\Phi^{\text{mag}}}{dt} \qquad \text{(Faraday's law).} \qquad \text{(Eq. 31-21)}$$

We owe to Faraday and Maxwell the discovery that a changing magnetic field produces an electric field. We owe to Maxwell alone the symmetric prediction that a changing electric field ought to produce a magnetic field. Maxwell made this prediction by noting a mathematical similarity between Ampère's law, which describes the nature of the magnetic field generated by current flow in a conductor

$$\oint \vec{B} \cdot d\vec{s} = \mu_0 i^{\text{enc}} \qquad \text{(Ampère's law),} \qquad \text{(Eq. 30-16)}$$

and the general formulation of Faraday's law

$$\mathcal{E} = \oint \vec{E} \cdot d\vec{s} = -\frac{d\Phi^{\text{mag}}}{dt} \qquad \text{(Faraday's law).} \qquad \text{(Eq. 31-21)}$$

At this point Maxwell speculated that the magnetic field induced along a closed loop (Ampère's law) was more generally a function of the rate of change of the electric flux enclosed by the loop. He predicted that any region in space with changing electric flux (like that found between plates when a capacitor is charged or discharged) could induce a magnetic field in the region—even without the presence of a conductor. This notion prompted Maxwell in 1861 to invent the concept of displacement current (discussed in Section 31-9). This fictitious current was devised to describe the possible, but as yet unobserved, magnetic effects of changing electric flux. Maxwell incorporated his displacement current concept into the reformulation of Ampère's law that we presented in Section 31-8,

$$\oint \vec{B} \cdot d\vec{s} = \mu_0 \varepsilon_0 \frac{d\Phi^{\text{elec}}}{dt} + \mu_0 i^{\text{enc}} \qquad \text{(Ampère–Maxwell law).} \qquad \text{(Eq. 31-32)}$$

Since there was no experimental evidence that changing electric flux could generate a magnetic field in the absence of real current, the proposed Ampère–Maxwell law was perhaps the most remarkable of Maxwell's many predictions.

Note that in free space, with no conductors present, the $\mu_0 i$ term in the expression above (Eq. 31-32) disappears and it becomes symmetric to Faraday's law (Eq. 31-21). The symmetry between these equations provided the basis for Maxwell's belief in the existence of electromagnetic waves almost 25 years before Hertz was able to generate and detect them. To find out why, read on.

Electromagnetic Wave Propagation

Maxwell used the symmetric relationship between Faraday's law and the Ampère–Maxwell law to predict a phenomenon that at first glance seems quite bizarre. He noted that if a changing magnetic field could create a changing electric field, then the changing electric field could, in turn, create another changing magnetic field. These changing fields could continuously generate each other and propagate, carrying "electromagnetic" energy with them.

In his 1861 paper (where he also introduced the displacement current concept) Maxwell rewrote the Faraday and Ampère–Maxwell laws as differential equations and solved them to describe $\vec{E}$ and $\vec{B}$ fields separately as functions of time. This produced differential equations that have the same algebraic form as those that describe the propagation of pressure variations in air (sound waves), ripples on the surface of a pond (water waves), and transverse displacements of a stretched string. Hence, the concept of an "electromagnetic wave" was born.

A rigorous treatment of Maxwell's mathematical description of electromagnetic waves would require us to solve differential equations and interpret the results. These methods are beyond the scope of this text. Instead we present some new results that arise from Maxwell's equations and show that they are consistent with experimental results. This will provide us with insight into the generation and properties of electromagnetic radiation. For example, one outcome of solving Maxwell's wave equations was the revelation that an electromagnetic wave ought to travel in a vacuum at a speed given by

$$c = \frac{1}{\sqrt{\mu_0 \varepsilon_0}} \qquad \text{(predicted wave speed in a vacuum).} \qquad (34\text{-}1)$$

Maxwell was quite surprised to find that the wave speed depends on the familiar electric and magnetic constants, ε_0 and μ_0, previously measured in static, electric, and magnetic experiments. He was equally surprised to find that a calculation of the speed in a vacuum agreed well with what was known at the time to be the speed of light. This led Maxwell to make the additional, rather bold, hypothesis that light was an electromagnetic wave.

34-3 The Generation of Electromagnetic Waves

In this section we give a more detailed description of how electromagnetic waves are generated. Observations summarized in Maxwell's equations reveal that electromagnetic waves (sometimes called "radiation") should be generated by accelerating charges. We begin by considering the generation of radio-frequency waves used for radio and TV transmission. These waves (wavelength $\lambda \approx 1$m) provide a source of radiation (the emitted waves) that is both macroscopic and of manageable dimensions so that classical physics rules. Some electromagnetic waves, including x-rays, gamma rays, and visible light, are *radiated* (emitted) from sources that are of atomic or nuclear size, where quantum physics rules.

At this point, you may find it helpful to quickly review mechanical waves discussed in Chapter 17. It was there that we first introduced important and pertinent wave-related concepts such as wavelength and frequency. We begin this section by considering the analogy between the generation and propagation of an electromagnetic *wave pulse* and a pulse traveling on a stretched string (Section 17-3) or along the surface of a pond (Section 18-1). We end the section with a discussion of how sinusoidal radio-frequency electromagnetic waves are generated.

An Electromagnetic Wave Pulse

How does a wave pulse propagate? Let's consider a more familiar situation in which a small bucket of water is dumped onto the surface of a pond. The water that is dumped on the pond's surface will undergo a rapid oscillation as it falls and rises again. However, the water that is dumped on the surface cannot undergo this oscillation without causing the ring of water that surrounds it to begin oscillating. This oscillation is passed along to the next ring of surrounding water, and so on. We see a two-dimensional wave crest like that shown in Fig. 18-3 traveling along the pond's surface at a constant speed.

In 1842, prior to Maxwell's prediction of electromagnetic waves, Joseph Henry observed electrical oscillations produced by a spark discharge from a capacitor. Because the theoretical basis for explaining electrical oscillations was not yet developed, the significance of these oscillations was not appreciated. Hertz generated electrical oscillations once again in 1888 using an induction coil to create sparks.

In order to better understand the connection between electromagnetic waves, electric oscillations, and the spark discharges with which Henry and Hertz experimented, we need to pull together ideas from other chapters. First, we know from

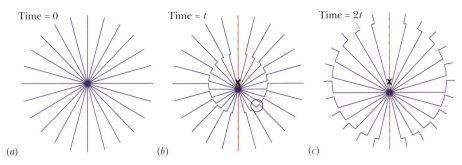

Time = 0   (a)   Time = t   (b)   Time = 2t   (c)

FIGURE 34-2 ▪ (*a*) Electric field lines in the plane of a resting point charge at the exact moment the jerk starts. (*b*) After an elapsed time *t* the kink at the interface between the old set of lines and the set associated with the jerk has traveled a distance *ct*. The original location of the charge is marked with an X and the line of motion is shown with a red dashed line. The region within the small circle is shown enlarged in Fig. 34-3. (*c*) After a time 2*t*, the kink is at a distance 2*ct*. (Diagrams adapted from *Electric & Magnetic Interactions: The Movies*, © 1996 Ruth Chabay and Bruce Sherwood, Carnegie Mellon University.)

Section 23-10 that electric field arrows indicate the direction of the electric field. Second, although we do not discuss this until Section 38-3, Einstein's principle of relativity establishes that nothing can travel faster than the speed of light. This includes information about the position of a charge. Keeping these ideas in mind, we can visualize how the field lines from a charge must readjust when the charge undergoes a sudden and brief acceleration like the spark discharge with which Henry and Hertz experimented. Specifically, suppose that at a time $t_1 = 0$ s a point charge initially at rest is suddenly accelerated straight downward and then stopped, all in an extremely short time period Δt. This is shown in Fig. 34-2. The original location of the charge is marked with an "X" in Fig. 34-2*b* and *c*.

What happens? As discussed below Maxwell's equations would lead us to predict that the accelerating charge will generate a three-dimensional electromagnetic wave crest. Furthermore, if you consider how this situation is like the momentary acceleration of the bit of mass at the end of a taut string when it is given a quick downward jerk, the generation of a pulse makes some intuitive sense as well. Let's look at the situation more carefully.

Before the acceleration, the electric field lines associated with the charge look just like those depicted Fig. 34-2*a*. Electromagnetic information travels at the speed of light, so after a time period $\Delta t = t$ the field lines will point to the charge's new position at the end of the acceleration — but only up to a distance *ct*. This is shown in Fig. 34-2*b*. Beyond that distance, the field lines will point to the charge's old position because the "news" of the charge's changed position will not have had time to spread that far.

At the distance *ct*, the field lines will have a kink joining the new set and the old set. This kink is increasingly close to perpendicular to the direction of motion of the wave front the farther you get from the accelerating charge. If we assume that field lines are continuous and that the field vectors are tangent to the line at each point, then the $\vec{E}$ field vectors in the kink are also increasingly perpendicular to the direction of motion of the wavefronts. The kink travels outward at the speed of light so that it has moved to a distance of 2*ct* after a time of $\Delta t = 2t$. This pulse continues to move out at speed *c* from the original location of the charge. Figure 34-3 shows a magnified view of a circled piece of Fig. 34-2*b* with extra field vectors drawn in.

In the brief moment that the charge is being accelerated, the "current" created by it is increasing ($di/dt = d^2q/dt^2 > 0$). We can use Ampère's law to determine the direction of the increasing magnetic field associated with this increasing "current." Taking the line of motion to be the direction of our "current" (shown with a red dashed line

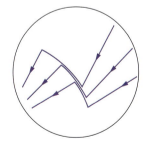

FIGURE 34-3 ▪ Here more $\vec{E}$ field vectors are assigned to the charge and shown in a magnified view of a circled piece of Fig. 34-2*b*. The crest of the wave pulse at time *t* show that the density of electric field lines is higher than the density of lines created when the charge is resting. Note that the $\vec{E}$ field in the kink points in a direction that is nearly perpendicular to the propagating electromagnetic wave pulse.

in Fig. 34-2*b*) and using Ampère's law, we see that the magnetic field points into the page to the left of the line of motion of the charge and out of the page to the right. The direction of these $\vec{B}$ field vectors (into or out of the page) is always perpendicular to that of the $\vec{E}$ field vectors. So, we surmise that the $\vec{B}$ field vectors associated with the kink will point perpendicular to the direction of motion of the kink. There are no propagating $\vec{E}$ or $\vec{B}$ fields along the line of motion of the charge. We define the wave disturbance as the propagation of the $\vec{E}$ field and $\vec{B}$ field vectors that are each perpendicular to the direction of motion of the wave. So, we have an electromagnetic wave pulse that is *transverse*. (See Chapter 17 for a review of wave properties if this last comment is not clear.)

Although this informal consideration of a single wave pulse created by a sudden acceleration of a charge is not rigorous or complete, it does give us a useful qualitative picture of how a pulsed electromagnetic wave might be generated.

Continuous Electromagnetic Wave Generation

How can we generate a continuous electromagnetic wave? What corresponds to the generation of the sinusoidal oscillation in a stretched string as shown in Fig. 17-6? Suppose we cause charges to oscillate back and forth along a line with sinusoidal or simple harmonic motion (SHM). We know from Section 16-4 that these charges will also have a sinusoidal acceleration, and so by the discussion above, we know that they will generate continuous electromagnetic (EM) waves.

Figure 34-4 is a schematic of an apparatus that can be used to generate radio frequency waves. The apparatus is an LC oscillator like that described in Section 33-6, but with a broadcast antenna coupled to it. This oscillator, mentioned in the previous sentence, establishes a sinusoidal current with an angular frequency given by $\omega = 1/\sqrt{LC}$. An ac generator provides a source of energy to compensate for thermal losses in the oscillator circuit and also for the energy carried away by the radiated electromagnetic wave.

The LC oscillator is coupled by a transformer and a transmission line to an *antenna*, which consists essentially of two thin, solid, conducting rods. Through this coupling, the sinusoidally varying current in the oscillator causes charge to oscillate sinusoidally along the antenna rods. The antenna behaves like an electric dipole described in Section 25-8, except that its electric dipole moment along the antenna changes sinusoidally in magnitude and direction over time. Since the charges are oscillating, they are continually accelerating and thereby produce electromagnetic radiation.

Because the dipole moment varies in the antenna in magnitude and direction, the electric and magnetic fields produced by the dipole vary in magnitude and direction. However, the changes in the electric and magnetic fields do not happen everywhere instantaneously. Rather, the changes travel outward from the antenna at the speed of light *c*. Together the changing fields form an electromagnetic wave that travels away from the antenna at speed *c*. The angular frequency of this wave, *ω*, is the same as that of the LC oscillator.

FIGURE 34-4 ■ Apparatus for generating a traveling electromagnetic wave at a "shortwave" radio frequency. An LC oscillator produces a sinusoidal current in an antenna. This generates an oscillating magnetic field and thus an EM wave. *P* is a distant point at which a detector (consisting of a dipole receiving antenna) could be placed to monitor the wave traveling past it.

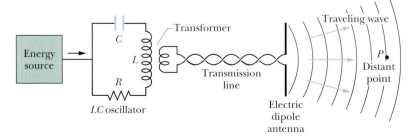

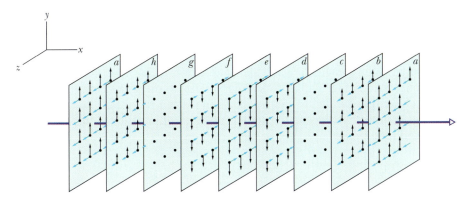

FIGURE 34-5 ■ A "snapshot" of some equally spaced planes of a sinusoidal electromagnetic wave traveling in the $+x$ direction. In this particular wave, the darker, vertically oriented $\vec{E}$ field vectors always point along the y axis. The lighter, horizontally oriented $\vec{B}$ field vectors always point along the z axis. The planes are labeled a through h to correspond to the $\vec{E}$ and $\vec{B}$ field vector configurations shown in Fig. 34-6.

The wave moves out in all directions except along the line of motion of the charges. In a direction perpendicular to the antenna the wavefronts will be approximately spherical in shape. If we go to a point that is far from the charge, the spherical wavefronts seem almost flat (just as the Earth seems flat to us because we are so far from its center). If we block all but a small piece of this wavefront, we get what looks like a series of planes marching forward at the speed of light. Hence, we often refer to electromagnetic waves as "plane waves." Although the plane wave approximation is a simplification of the real situation, it describes the nature of waves quite well when the distance from the source, in this case the antenna, is large compared to the wavelength of electromagnetic wave. The value to us in looking at electromagnetic waves as **"plane waves"** is that they are simpler to deal with mathematically than spherical waves.

There are several ways to depict a traveling sinusoidal electromagnetic plane wave. One of these is shown in Fig. 34-5. Here we show a "snapshot" of a few equally spaced planes of a sinusoidal electromagnetic wave traveling in the $+x$ direction. In this particular wave, the darker, vertically oriented $\vec{E}$ field vectors always point along the y axis. The lighter, horizontally oriented $\vec{B}$ field vectors always point along the z axis. This diagram emphasizes the fact that, at a given instant, field vectors are the same everywhere in a given y-z plane. The figure also shows that as the wave passes a point in space, the field vector values will sinusoidally vary from a maximum to a minimum and back again. This is shown by the dashed lines representing curves through the tips of the $\vec{E}$ vectors and the $\vec{B}$ vectors at the bottom of each plane. The planes are labeled a through h to correspond to the $\vec{E}$ and $\vec{B}$ field vector configurations shown in Fig. 34-6.

Figure 34-6 shows how the electric field $\vec{E}$ and the magnetic field $\vec{B}$ change with time as one wavelength of the wave sweeps past the distant point P of Fig. 34-4. In each part of Fig. 34-5, the wave is traveling directly along the x axis. There are several key features of any sinusoidal plane electromagnetic wave that are shown in Fig. 34-5 and Fig. 34-6 that are present whenever a plane electromagnetic wave is created with charges that oscillate sinusoidally along a line:

1. The electric and magnetic fields $\vec{E}$ and $\vec{B}$ are always perpendicular to the direction of travel of the wave. Thus, the wave is a *transverse wave*, as discussed in Chapter 17.

2. The electric field is always perpendicular to the magnetic field.

3. The cross product $\vec{E} \times \vec{B}$ always points in the direction of travel of the wave.

4. For a single simple plane wave, the fields always vary sinusoidally over time, just like the transverse waves discussed in Chapter 17. Moreover, the fields vary with the same frequency and are *in phase* (in step) with each other. More complex fields can be described mathematically as a superposition of plane waves.

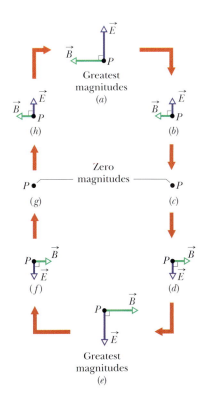

FIGURE 34-6 ■ (a)–(h) The variation in the electric field $\vec{E}$ and the magnetic field $\vec{B}$ at the distant point P of Fig. 34-4 as one wavelength of the electromagnetic wave shown in Fig. 34-5 travels past it. In this perspective, the wave is traveling directly out of the page. The two fields vary sinusoidally in magnitude and direction. Each of the planes that correspond to parts of this diagram (a)–(h) are also shown in Fig. 34-5.

34-4 Describing Electromagnetic Wave Properties Mathematically

Suppose an electromagnetic wave like that shown in Fig. 34-5 is traveling toward P, the electric field in Fig. 34-5 is oscillating parallel to the y axis, and the magnetic field is then oscillating parallel to the z axis. Then we can represent the electric and magnetic fields mathematically as sinusoidal functions of position x (along the path of the wave) and time t:

$$\vec{E} = \vec{E}^{\,max} \sin(kx - \omega t), \qquad (34\text{-}2)$$

and
$$\vec{B} = \vec{B}^{\,max} \sin(kx - \omega t), \qquad (34\text{-}3)$$

where $\vec{E}^{\,max}$ and $\vec{B}^{\,max}$ represent the amplitudes of the fields and, as in Chapter 17, ω and k are the angular frequency and wave number, respectively. From these equations, we note that each type of field forms its own wave. Equation 34-2 gives the *electric wave component* of the electromagnetic wave, and Eq. 34-3 gives the *magnetic wave component*. As we already realize from considering Maxwell's formulation, these two wave components cannot exist independently. The wave propagates because a changing electric field creates a changing magnetic field, which generates another changing electric field, and so on.

At this point it is useful to devise a second "snapshot" representation of the plane electromagnetic wave. This is shown in Fig. 34-7b. Instead of emphasizing the planar nature of a selected sample of wave fronts as we did in Fig. 34-5, we show just one vector that represents the length of electric and magnetic field vectors in each sample plane. The curves through the tips of the vectors display the sinusoidal nature of the oscillations described by $\vec{E} = \vec{E}^{\,max} \sin(kx - \omega t)$, and $\vec{B} = \vec{B}^{\,max} \sin(kx - \omega t)$, (Eqs. 34-2 and 34-3) above. The wave components $\vec{E}$ and $\vec{B}$ are depicted as in phase, perpendicular to each other, and perpendicular to the wave's direction of travel.

Figure 34-7a shows how the new representation is tied to the old one. Only three sample planes with $\vec{E}$ vectors along one line, separated by a half wavelength $\lambda/2(= \pi/k)$ of the wave, are shown.

In some cases waves are traveling in approximately the same direction and form a beam, such as a laser beam or a beam of radio waves. A beam can be represented with a "ray," which is just a line showing the direction of motion of the beam. This is also

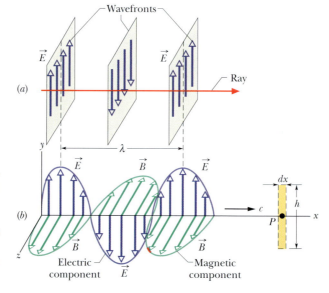

FIGURE 34-7 ■ (a) A plane electromagnetic wave represented with a ray and three wavefronts separated by a half wavelength $\lambda/2$. (b) The same wave represented in a "snapshot" of its electric field $\vec{E}$ and magnetic field $\vec{B}$ at points on the x axis, along which the wave travels at speed c. As it travels past point P, the fields vary as shown in Fig. 34-5. The dashed rectangle at P is used in Fig. 34-8a.

shown in Fig. 34-7*a*. "Rays" of light will become increasing prominent in our discussions of image formations in Chapters 35 and 36.

Interpretation of Fig. 34-7*b* requires some care. The similar drawings for a transverse wave on a taut string that we discussed in Chapter 17 represented the up and down displacement of sections of the string as the wave passed (*something actually moved*). Figure 34-7*b* is more abstract. At the instant shown, the electric and magnetic fields each have a certain magnitude and direction (but always perpendicular to the *x* axis) at each point along the *x* axis. We choose to represent these vector quantities with a pair of arrows for each point, so we must draw arrows of different lengths for different points, all directed away from the *x* axis, like thorns on a rose stem. For each line parallel to the *x* axis there is a similar picture. In viewing figures such as this, it is important to remember that the length of the arrows represents the field values along the line chosen as the *x* axis. Neither the arrows nor the sinusoidal curves represent a sideways displacement of anything.

A Most Curious Wave

From our previous work with waves in Chapters 17 and 18, we know that the speed of the wave is ω/k (Eq. 17-12). However, it is customary to use the symbol c (rather than v) to denote an electromagnetic wave speed in a vacuum (or air):

> All electromagnetic waves, including visible light, have the same speed c in a vacuum. Hence, c is called "the speed of light."*

The waves we discussed in Chapters 17 and 18 require a *medium* (some material) through which or along which to travel. We had waves traveling along a string, through the Earth, and through the air. Maxwell and other 19th-century investigators assumed there was a medium through which electromagnetic waves traveled. However, we now believe that electromagnetic waves are curiously different in that they require no medium for travel. They can, indeed, travel through a medium such as air or glass, but they can also travel through the vacuum of space between a distant star and the Earth.

Once Albert Einstein proposed the special theory of relativity in 1905, scientists realized that visible light waves and other electromagnetic waves were special entities. The reason is that light has the same speed in any inertial frame of reference. If you send a beam of light along an axis and ask several observers to measure its speed while they move at different speeds along that axis, either in the direction of the light or opposite it, the observers will all measure the *same speed* for the light. This result is an amazing one and quite different from what would have been found if those observers had measured the speed of any other type of wave. For other waves, the speed of the observers relative to the wave would have affected their measurements. The implications of this are striking and include some seemingly bizarre effects that we will learn about in Chapter 38 on special relativity.

As we saw in Chapter 1, the speed of light (any electromagnetic wave) in vacuum has the exact value

$$c = 299\ 792\ 458 \text{ m/s.}^{\dagger}$$

The Ratio of $\vec{E}$ to $\vec{B}$ and the Induced Electric Field

We can use Faraday's law

$$\mathcal{E} = \oint \vec{E} \cdot d\vec{s} = -\frac{d\Phi^{\text{mag}}}{dt}$$

---

* The letter c comes from the Latin word *celar,* which means "fast".

† The value is "exact" so the meter could be redefined as a length of path traveled by light in a specified time. (See Section 1-6 for details.)

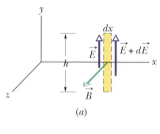

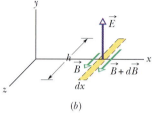

FIGURE 34-8 ■ (*a*) As the electromagnetic wave travels rightward past point *P* in Fig. 34-7, the sinusoidal variation of the magnetic field $\vec{B}$ through a rectangle centered at *P* induces electric fields along the rectangle. At the instant shown, $\vec{B}$ is decreasing in magnitude and the induced electric field is therefore greater in magnitude on the right side of the rectangle than on the left. (*b*) The sinusoidal variation of the electric field through this rectangle, located (but not shown) at point *P* in Fig. 34-8*a*, induces magnetic fields along the rectangle. The instant shown is that of Fig. 34-8*a*: $\vec{E}$ is decreasing in magnitude, and the induced magnetic field is greater in magnitude on the right side of the rectangle than on the left.

to find the ratio of the electric and magnetic fields at any location along an electromagnetic wave. We start by considering the dashed rectangle (of dimensions *dx* and *h* in Fig. 34-7*b*) that is fixed at point *P* on the *x* axis and in the *xy* plane. As the electromagnetic wave moves to the right past the rectangle, the magnetic flux Φ^{mag} through the rectangle changes and—according to Faraday's law of induction—induced electric fields appear throughout the region of the rectangle. We take $\vec{E}$ and $\vec{E} + d\vec{E}$ to be the induced fields along the two long sides of the rectangle. These induced electric fields are, in fact, the electric component of the electromagnetic wave at those points.

Let us consider these fields at the instant when the magnetic wave component passing through the rectangle is the small section marked with red on the line marked "a magnetic field component" in Fig. 34-7*b*. Just then, the magnetic field through the rectangle points in the positive *z* direction and is decreasing in magnitude (the magnitude was greater just before the red section arrived). Because the magnetic field is out of the page and decreasing, the magnetic flux Φ^{mag} through the rectangle is also decreasing. According to Faraday's law, this change in flux is opposed by induced electric fields. This implies that a counterclockwise induced current would appear along the rectangle if it were a conductor (which it is not). This in turn implies that a counterclockwise induced electric field would have to appear along the rectangle. So, the induced electric field vectors $\vec{E}$ and $\vec{E} + d\vec{E}$ are indeed oriented as shown in Fig. 34-8*a*, with the magnitude of $\vec{E} + d\vec{E}$ greater than that of $\vec{E}$. Otherwise, the net induced electric field would not act counterclockwise around the rectangle.

Let us now apply Faraday's law of induction,

$$\oint \vec{E} \cdot d\vec{s} = -\frac{d\Phi^{mag}}{dt}, \tag{34-4}$$

proceeding counterclockwise around the rectangle of Fig. 34-8*a*. There is no contribution to the integral from the top or bottom of the rectangle because $\vec{E}$ and $d\vec{s}$ are perpendicular there. The integral then has the value

$$\oint \vec{E} \cdot d\vec{s} = (|\vec{E}| + d|\vec{E}|)h - |\vec{E}|h = h\,d|\vec{E}|. \tag{34-5}$$

The flux Φ^{mag} through this rectangle is

$$\Phi^{mag} = |\vec{B}|h\,dx, \tag{34-6}$$

where $|\vec{B}|$ is the magnitude of $\vec{B}$ within the rectangle and *h dx* is the area of the rectangle. Differentiating this expression (Eq. 34-6) with respect to *t* gives

$$\frac{d\Phi^{mag}}{dt} = h\,dx\frac{d|\vec{B}|}{dt}. \tag{34-7}$$

If we substitute this result (Eq. 34-7) and $\oint \vec{E} \cdot d\vec{s} = h\,d|\vec{E}|$ (Eq. 34-5) into Faraday's law $\oint \vec{E} \cdot d\vec{s} = -d\Phi^{mag}/dt$ (Eq. 34-4), we find

$$h\,d|\vec{E}| = -h\,dx\frac{d|\vec{B}|}{dt}$$

or

$$\frac{d|\vec{E}|}{dx} = -\frac{d|\vec{B}|}{dt}. \tag{34-8}$$

We gather from our electromagnetic wave representations and from $\vec{E} = \vec{E}^{max}\sin(kx - \omega t)$, and $\vec{B} = \vec{B}^{max}\sin(kx - \omega t)$ (Eqs. 34-2 and 34-3) that both $|\vec{B}|$

and $|\vec{E}|$ are functions of two variables, x and t. However, in evaluating $d|\vec{E}|/dx$, we can assume that t is constant because we consider only an "instantaneous snapshot." Also, in evaluating $d|\vec{B}|/dt$ we can assume that x is constant because we are dealing with the time rate of change of $|\vec{B}|$ at a particular place, the point P in Fig. 34-7b. The derivatives under these circumstances are partial derivatives, and Eq. 34-8 must be written

$$\frac{\partial|\vec{E}|}{\partial x} = -\frac{\partial|\vec{B}|}{\partial t}. \tag{34-9}$$

The minus sign in this equation is appropriate and necessary because, although $|\vec{E}|$ is increasing with x at the site of the rectangle in Fig. 34-8, $|\vec{B}|$ is decreasing with t.

From $\vec{E} = \vec{E}^{\,\text{max}}\sin(kx - \omega t)$ (Eq. 34-2) we have

$$\frac{\partial\vec{E}}{\partial x} = k\vec{E}^{\,\text{max}}\cos(kx - \omega t)$$

and from $\vec{B} = \vec{B}^{\text{max}}\sin(kx - \omega t)$ (Eq. 34-3)

$$\frac{\partial\vec{B}}{\partial t} = -\omega\vec{B}^{\,\text{max}}\cos(kx - \omega t).$$

Then $\partial|\vec{E}|/\partial x = -\partial|\vec{B}|/\partial t$ (Eq. 34-9) reduces to

$$k|\vec{E}^{\,\text{max}}|\cos(kx - \omega t) = \omega|\vec{B}^{\,\text{max}}|\cos(kx - \omega t). \tag{34-10}$$

The ratio ω/k for a traveling wave is its speed, which we are calling c. Hence, we see that

$$\frac{|\vec{E}^{\,\text{max}}|}{|\vec{B}^{\,\text{max}}|} = c \qquad \text{(amplitude ratio)}. \tag{34-11}$$

If we divide $\vec{E} = \vec{E}^{\text{max}}\sin(kx - \omega t)$ by $\vec{B} = \vec{B}^{\,\text{max}}\sin(kx - \omega t)$ (Eqs. 34-2 and 34-3) and then substitute into Eq. 34-10, we find that the ratio of magnitudes of the fields at every instant is given by

$$\frac{|\vec{E}|}{|\vec{B}|} = c \qquad \text{(magnitude ratio)}. \tag{34-12}$$

Induced Magnetic Field and the Equation for Wave Speed

If we use the Ampère–Maxwell law

$$\oint \vec{B}\cdot d\vec{s} = \mu_0\varepsilon_0\frac{d\Phi^{\text{elec}}}{dt} + \mu_0 i^{\text{enc}},$$

we can find an alternative expression for the wave speed in the case where no real current is present (so $i^{\text{enc}} = 0$ A). We start with Fig. 34-8b, which shows another dashed rectangle at point P of Fig. 34-7; this one is in the xz plane. As the electromagnetic wave moves rightward past this new rectangle, the electric flux Φ^{elec} through the rectangle changes and—according to the Ampère–Maxwell law of induction—induced magnetic fields appear throughout the region of the rectangle. These induced magnetic fields are, in fact, the magnetic field components of the electromagnetic wave.

We see from Fig. 34-7 that at the instant chosen for the magnetic field in Fig. 34-8, the electric field through the rectangle of Fig. 34-8b is directed as shown. Recall that at the chosen instant, the magnetic field in Fig. 34-8a is decreasing. Because the two fields are in phase, the electric field in Fig. 34-8b must also be decreasing, and so must the electric flux Φ^{elec} through the rectangle. By applying the same reasoning we applied to Fig. 34-8a, we see that the changing flux Φ^{elec} will induce a magnetic field with vectors $\vec{B}$ and $\vec{B} + d\vec{B}$ oriented as shown in Fig. 34-8b, where $\vec{B} + d\vec{B}$ is greater than $\vec{B}$.

Let us also apply Maxwell's law of induction with no real current present,

$$\oint \vec{B} \cdot d\vec{s} = \mu_0 \varepsilon_0 \frac{d\Phi^{elec}}{dt}, \tag{34-13}$$

by proceeding counterclockwise around the dashed rectangle of Fig. 34-8b. Only the long sides of the rectangle contribute to the integral, whose value is

$$\oint \vec{B} \cdot d\vec{s} = -(|\vec{B}| + d|\vec{B}|)h + |\vec{B}|h = -h\,d|\vec{B}|. \tag{34-14}$$

The flux Φ^{elec} through the rectangle is

$$\Phi^{elec} = (|\vec{E}|)(h\,dx), \tag{34-15}$$

where $|\vec{E}|$ is the average magnitude of $\vec{E}$ within the rectangle. Differentiating this expression with respect to t gives

$$\frac{d\Phi^{elec}}{dt} = h\,dx\frac{d|\vec{E}|}{dt}.$$

If we substitute this and $\oint \vec{B} \cdot d\vec{s} = -h\,d|\vec{B}|$ (Eq. 34-14 from above) into Maxwell's law of induction we find that

$$-h\,d|\vec{B}| = \mu_0 \varepsilon_0 \left(h\,dx\frac{d|\vec{E}|}{dt} \right).$$

Changing to partial-derivative notation as we did before (Eq. 34-9),

$$-\frac{\partial|\vec{B}|}{\partial x} = \mu_0 \varepsilon_0 \frac{\partial|\vec{E}|}{\partial t}. \tag{34-16}$$

Again, the minus sign in this equation makes sense because, although $\vec{B}$ is increasing with x at point P in the rectangle in Fig. 34-8b, $\vec{E}$ is decreasing with t.

Evaluating this expression by using $\vec{E} = \vec{E}^{max} \sin(kx - \omega t)$, and $\vec{B} = \vec{B}^{max} \sin(kx - \omega t)$ (Eqs. 34-2 and 34-3) leads to

$$-k|\vec{B}^{max}|\cos(kx - \omega t) = -\mu_0 \varepsilon_0 \omega|\vec{E}^{max}|\cos(kx - \omega t),$$

which we can write as

$$\frac{|\vec{E}^{max}|}{|\vec{B}^{max}|} = \frac{1}{\mu_0 \varepsilon_0 (\omega/k)} = \frac{1}{\mu_0 \varepsilon_0 c}.$$

Combining this with $|\vec{E}^{max}|/|\vec{B}^{max}| = c$ (Eq. 34-11) leads at once to

$$c = \frac{1}{\sqrt{\mu_0 \varepsilon_0}} \quad \text{(wave speed),} \tag{34-17}$$

which is exactly Eq. 34-1.

READING EXERCISE 34-1: The magnetic field $\vec{B}$ through the rectangle of Fig. 34-8a is shown at a different instant in part 1 of the accompanying figure; $\vec{B}$ is directed in the xz plane, parallel to the z axis, and its magnitude is increasing. (a) Complete part 1 by drawing the induced electric fields, indicating both directions and relative magnitudes (as in Fig. 34-8a). (b) For the same instant, complete part 2 of the figure by drawing the electric field of the electromagnetic wave. Also draw the induced magnetic fields, indicating both directions and relative magnitudes (as in Fig. 34-8b).

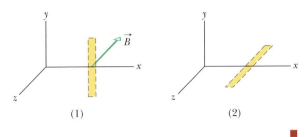

(1)   (2)

34-5 Transporting Energy with Electromagnetic Waves

From our experience with capacitors, we know that energy is stored in an electric field. Likewise, from our experience with inductors we know that energy is stored in a magnetic field. So, it makes sense that as an electromagnetic wave moves through space, it carries energy with it. Sunbathers will confirm this hunch. An electromagnetic wave can transport energy and deliver it to a body on which it falls. In this section, we develop an expression that will allow us to quantify the rate per unit area at which energy is transported by an electromagnetic wave. This quantity is a measure of the **intensity** of the electromagnetic wave.

From Chapter 28 on capacitors, we know that the energy per unit area or *energy density* within an electric field is

$$u^{\text{elec}} = \tfrac{1}{2}\varepsilon_0 E^2.$$

Because you also know (from this chapter) that $\vec{E} = c\vec{B}$ and c is such a very large number, you might conclude that the energy associated with the electric field in an electromagnetic wave is much greater than that associated with the magnetic field. That conclusion is incorrect; the densities of the two energies are exactly equal. To show this, we substitute $c\vec{B}$ for $\vec{E}$; then we can write

$$u^{\text{elec}} = \tfrac{1}{2}\varepsilon_0 E^2 = \tfrac{1}{2}\varepsilon_0 c^2 B^2.$$

If we now substitute for c with (Eq. 34-1)

$$c = \frac{1}{\sqrt{\varepsilon_0 \mu_0}},$$

we get

$$u^{\text{elec}} = \tfrac{1}{2}\varepsilon_0 \frac{1}{\mu_0 \varepsilon_0} B^2 = \frac{B^2}{2\mu_0}.$$

In Section 33-3, we found that $B^2/2\mu_0$ is the energy density u^{mag} of a magnetic field $\vec{B}$. So, we see that $u^{\text{elec}} = u^{\text{mag}}$ everywhere along an electromagnetic wave.

> The energy density of the electric field is equal to the energy density of the magnetic field at every instant and for every point along an electromagnetic wave.

The total energy density for the electromagnetic wave is the sum of the energy density associated with the magnet field and the energy density associated with the electric field:

$$u^{\text{tot}} = u^{\text{elec}} + u^{\text{mag}} = \tfrac{1}{2}\varepsilon_0 E^2 + \frac{1}{2\mu_0}B^2.$$

However, since these values are equal to one another, we can also write the total energy density as twice either value:

$$u^{\text{tot}} = 2u^{\text{elec}} = \varepsilon_0 E^2$$

or

$$u^{\text{tot}} = 2u^{\text{mag}} = \frac{1}{\mu_0}B^2.$$

This result is quite helpful in developing an expression for the rate of energy (or power) transport across a unit of area perpendicular to the direction of propagation of the electromagnetic wave. At any instant, the **rate of energy transport per unit area** is

$$S = \left(\frac{\text{energy/time}}{\text{area}}\right)^{\text{inst}} = \left(\frac{\text{power}}{\text{area}}\right)^{\text{inst}} \quad \text{(instantaneous).} \qquad (34\text{-}18)$$

Note that from this we can see that the SI unit for S must be the watt per square meter (W/m^2). Since an electromagnetic wave moves with a speed c, in a time period Δt, the wave travels a distance $c\,\Delta t$. During that motion, if it passes through a surface of some area A, the volume of space through which the wave passes is $c\,\Delta t A$. The total energy transported by the wave is the total energy density (electric and magnetic) multiplied by this volume. That is, the total energy (U) transported by the wave to an area A in a time period Δt is

$$U = u^{\text{tot}}c\,\Delta t A,$$

where

$$u^{\text{tot}} = \varepsilon_0 E^2.$$

Hence,

$$U = \varepsilon_0 E^2 c\,\Delta t A.$$

As expressed by Eq. 34-18 above, the rate of energy transport is this total energy transported divided by the area through which the wave travels and the time period. That is,

$$S = \frac{\varepsilon_0 E^2 c\,\Delta t A}{\Delta t A} = \varepsilon_0 E^2 c.$$

Since $c^2 = 1/\varepsilon_0\mu_0$, we can multiply this expression by one in the form of $c^2\varepsilon_0\mu_0$. This gives an alternative expression for the instantaneous energy flow rate:

$$S = \frac{1}{c\mu_0}E^2 = \varepsilon_0 c\, E^2 \quad \text{(instantaneous energy flow rate).} \qquad (34\text{-}19)$$

By substituting $\vec{E} = \vec{E}^{\text{max}}\sin(kx - \omega t)$ into Eq. 34-19, we could obtain an equation for the energy transport rate as a function of time. More useful in practice, however, is the average energy transported over time. For that, we need to find the time-averaged value of S, written $\langle S \rangle$. We call this quantity the **intensity** I of the wave. Thus the intensity I is

$$I = \langle S \rangle = \left\langle \frac{\text{energy/time}}{\text{area}} \right\rangle = \left\langle \frac{\text{power}}{\text{area}} \right\rangle. \tag{34-20}$$

With $S = E^2/c\mu_0$ (Eq. 34-19), we find

$$I = \langle S \rangle = \frac{1}{c\mu_0}\langle E^2 \rangle = \frac{1}{c\mu_0}\langle (E^{\max})^2 \sin^2(kx - \omega t) \rangle. \tag{34-21}$$

Over a full cycle, the average value of $\sin^2 \theta$, for any angular variable θ, is $\frac{1}{2}$. In addition, we define a new quantity E^{rms}, the root-mean-square value of the electric field magnitude as

$$E^{\text{rms}} \equiv \frac{E^{\max}}{\sqrt{2}}. \tag{34-22}$$

We can then rewrite Eq. 34-21 as

$$I = \frac{1}{c\mu_0}(E^{\text{rms}})^2. \tag{34-23}$$

If we combine the ideas that we have developed above for the rate of energy transported by an electromagnetic wave per unit area with knowledge of the direction in which the wave is traveling, we can define a vector $\vec{S}$ that describes both the energy transport rate S *and* the direction in which the transfer in occurring. This vector is an important quantity, so we give it a name. It is called the **Poynting vector** after John Henry Poynting (1852–1914), who first discussed its properties.

The direction of the Poynting vector $\vec{S}$ of an electromagnetic wave at any point gives the wave's direction of travel and so the direction of energy transport at that point.

We can combine everything that we have developed in this section into a single expression defining the Poynting vector $\vec{S}$ as

$$\vec{S} = \frac{1}{\mu_0}\vec{E} \times \vec{B} \qquad \text{(instantaneous Poynting vector)}, \tag{34-24}$$

where magnitude $|\vec{S}|$ of the Poynting vector is the instantaneous intensity of the electromagnetic wave. $\vec{E}$ and $\vec{B}$ are perpendicular to each other in an electromagnetic wave. Hence, the magnitude of $\vec{E} \times \vec{B}$ is $|\vec{E}||\vec{B}|$. Since $|\vec{E}| = c|\vec{B}|$, this is consistent with the expression that we developed above,

$$S = |\vec{S}| = \frac{1}{c\mu_0}E^2,$$

for the magnitude of energy transport rate. You can confirm for yourself that the cross product gives the correct direction for the wave propagation, and hence for the energy transport.

Variation of Intensity with Distance

The intensity variation of electromagnetic radiation with distance is often complex—especially when the source beams the radiation in a particular direction (like a searchlight at a movie premiere). We know that as a wavefront spreads out over a wider surface, its energy density must diminish, but how? Let's consider the simplest case we can imagine. Assume that the source is a *point source* that emits the light **isotropically**—

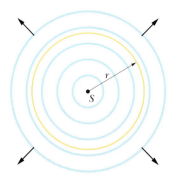

FIGURE 34-9 ■ A point source S emits electromagnetic waves uniformly in all directions. The spherical wavefronts pass through an imaginary sphere of radius r that is centered on S.

that is, with equal intensity in all directions. The spherical wavefronts spreading from such an isotropic point source S at a particular instant are shown in cross section in Fig. 34-9.

In a vacuum there is no mechanism for dissipation of energy, so it is conserved. Let us center an imaginary sphere of radius r on the source, as shown in Fig. 34-9. All the energy emitted by the source must pass through the sphere. Thus, the rate at which energy is transferred through the sphere by the radiation must equal the rate at which energy is emitted by the source—that is, the power P_s of the source. The intensity I (= power/area) at the sphere must then be

$$I = \frac{P_s}{4\pi r^2},\qquad(34\text{-}25)$$

where $4\pi r^2$ is the area of the sphere. This expression tells us that the intensity of the electromagnetic radiation from an isotropic point source decreases with the square of the distance r from the source.

Since from Eq. 34-23 we also know

$$I = \frac{1}{c\mu_0}(E^{\text{rms}})^2,$$

we can equate these two expressions to give

$$\frac{P_s}{4\pi r^2} = \frac{(E^{\text{rms}})^2}{c\mu_0},$$

or

$$(E^{\text{rms}})^2 = \frac{c\mu_0 P_s}{4\pi r^2}.$$

Simplification gives the relationship between the average electric field E^{rms} for a radiating charge, the power of the source P_s and the distance from the source r:

$$E^{\text{rms}} = \sqrt{\frac{c\mu_0 P_s}{4\pi}}\,\frac{1}{r}.\qquad(34\text{-}26)$$

This equation tells us that the electric field associated with an isotropic radiation point source falls off as $1/r$, rather than as $1/r^2$ as it does for a static electric field.

READING EXERCISE 34-2: The figure gives the electric field of an electromagnetic wave at a certain point and a certain instant. The wave is transporting energy in the negative z direction. What is the direction of the magnetic field of the wave at that point and instant?

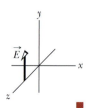

TOUCHSTONE EXAMPLE 34-1: Isotropic Light Source

As we stated in the introduction, visible light is now known to consist of electromagnetic waves. An observer is 1.8 m from an isotropic point light source whose power P_s is 250 W. Calculate the rms values of the electric and magnetic fields due to the source at the position of the observer.

SOLUTION ■ The first two **Key Ideas** here are these:

1.   The rms value E^{rms} of the electric field in light is related to the intensity I of the light by $I = (E^{\text{rms}})^2/c\mu_0$.

2.   Because the source is a point source emitting light with equal intensity in all directions, the intensity I at any distance r from the source is related to the source's power P_s via Eq. 34-25 ($I = P_s/4\pi r^2$).

Putting these two ideas together gives us

$$I = \frac{P_s}{4\pi r^2} = \frac{(E^{rms})^2}{c\mu_0},$$

which leads to

$$E^{rms} = \sqrt{\frac{P_s c\mu_0}{4\pi r^2}}$$

$$= \sqrt{\frac{(250 \text{ W})(3.00 \times 10^8 \text{ m/s})(4\pi \times 10^{-7} \text{ H/m})}{(4\pi)(1.8 \text{ m})^2}}$$

$$= 48.1 \text{ V/m} \approx 48 \text{ V/m}. \qquad \text{(Answer)}$$

The third **Key Idea** here is that magnitudes of the electric field and magnetic field of an electromagnetic wave at any instant and at any point in the wave are related by the speed of light c

according to Eq. 34-12 ($|\vec{E}|/|\vec{B}| = c$). Thus, the rms values of those fields are also related by Eq. 34-12 and we can write

$$B^{rms} = \frac{E^{rms}}{c}$$

$$= \frac{48.1 \text{ V/m}}{3.00 \times 10^8 \text{ m/s}}$$

$$= 1.6 \times 10^{-7} \text{ T}. \qquad \text{(Answer)}$$

Note that E^{rms} ($= 48$ V/m) is appreciable as judged by ordinary laboratory standards, but $B^{rms}(=1.6 \times 10^{-7}$ T) is quite small. This difference helps to explain why most instruments used for the detection and measurement of electromagnetic waves are designed to respond to the electric component of the wave. It is wrong, however, to say that the electric component of an electromagnetic wave is "stronger" than the magnetic component. You cannot compare quantities that are measured in different units. As we have seen, the electric and magnetic components are on an equal basis as far as the propagation of the wave is concerned, because their average energies, which can be compared, are exactly equal.

34-6   Radiation Pressure

Electromagnetic waves carry linear momentum as well as energy. This means that we can exert a pressure—a **radiation pressure**—on an object by shining light on it. However, the pressure must be very small because, for example, you do not feel a camera flash pushing on you when it is used to take your photograph.

To see how Maxwell related radiation pressure to light intensity, let us shine a beam of electromagnetic radiation—visible light, for example—on an object for a time interval Δt. Further, let us assume that the object is free to move and that the radiation is entirely **absorbed** (taken up) by the object. This means that during the interval Δt, the object gains an energy ΔU from the radiation. Maxwell showed that the object also gains linear momentum. As usual we can represent momentum with a lowercase p and an object's momentum change as $\Delta\vec{p} = \vec{p}_2 - \vec{p}_1$. The magnitude $|\Delta\vec{p}|$ of the momentum change of the object is related to the energy change ΔU by

$$|\Delta\vec{p}| = \frac{\Delta U}{c} \qquad \text{(total absorption)}, \qquad (34\text{-}27)$$

where c is the speed of light. The direction of the momentum change of the object is the direction of the incident (incoming) beam that the object absorbs.

Instead of being absorbed, the radiation can be reflected by the object; that is, the radiation can be sent off in a new direction as if it bounced off the object. If the radiation is entirely reflected back along its original path, the magnitude of the momentum change of the object is twice that given above, or

$$|\Delta\vec{p}| = \frac{2\,\Delta U}{c} \qquad \text{(total reflection back along path)}. \qquad (34\text{-}28)$$

In the same way, an object undergoes twice as much momentum change when a perfectly elastic tennis ball is bounced from it as when it is struck by a perfectly inelastic ball (a lump of wet putty, say) of the same mass and velocity. If the incident radiation is partly absorbed and partly reflected, the momentum change of the object is between $\Delta U/c$ and $2\,\Delta U/c$.

From Newton's Second Law, we know that a change in momentum is related to a force by

$$\vec{F} = \frac{|\Delta\vec{p}|}{\Delta t}.\tag{34-29}$$

To find expressions for the force exerted by radiation in terms of the intensity I of the radiation, suppose that a flat surface of area A, perpendicular to the path of the radiation, intercepts the radiation. In time interval Δt, the energy intercepted by area A is

$$\Delta U = IA\,\Delta t.\tag{34-30}$$

If the energy is completely absorbed, then Eq. 34-27 tells us that $|\Delta\vec{p}| = IA\,\Delta t/c$. Then with $\vec{F} = |\Delta\vec{p}|/\Delta t$ (Eq. 34-29), the magnitude of the force on the area A is

$$|\vec{F}| = \frac{IA}{c} \qquad \text{(total absorption).}\tag{34-31}$$

Similarly, if the radiation is totally reflected back along its original path, Eq. 34-28 tells us that $|\Delta\vec{p}| = 2IA\,\Delta t/c$ and, from Eq. 34-29,

$$|\vec{F}| = \frac{2IA}{c} \qquad \text{(total reflection back along path).}\tag{34-32}$$

If the radiation is partly absorbed and partly reflected, the magnitude of the force on area A is between the values of IA/c and $2IA/c$.

The force per unit area on an object due to radiation is the radiation pressure P. (Note that we represent pressure, as usual, with a capital P.) We can find the pressure for total absorption and total reflection by dividing both sides of each equation (34-31 and 34-32) by A. We obtain

$$P^{\text{absorp}} = \frac{I}{c} \qquad \text{(pressure for total wave absorption),}\tag{34-33}$$

$$\text{and} \qquad P^{\text{refl}} = \frac{2I}{c} \qquad \text{(pressure for total wave reflection back along path).}\tag{34-34}$$

Just as with fluid pressure in Chapter 15, the SI unit of radiation pressure is the pascal (Pa) which equals a newton per square meter (N/m^2).

The development of laser technology has permitted researchers to achieve radiation pressures much greater than, say, that due to a camera flashlamp. This comes about because a beam of laser light—unlike a beam of light from a small lamp filament—can be focused to a tiny spot only a few wavelengths in diameter. This permits the delivery of great amounts of energy and momentum to small objects placed at that spot.

READING EXERCISE 34-3: Light of uniform intensity shines perpendicularly on a totally absorbing surface, fully illuminating the surface. If the area of the surface is decreased, do (a) the radiation pressure and (b) the radiation force on the surface increase, decrease, or stay the same? ■

TOUCHSTONE EXAMPLE 34-2: Comet Dust

When dust is released by a comet, it does not continue along the comet's orbit because radiation pressure from sunlight pushes it radially outward from the Sun. Assume that a dust particle is spherical with radius R, has density $\rho = 3.5 \times 10^3 \text{ kg/m}^3$, and totally absorbs the sunlight it intercepts. For what value of R does the gravitational force $\vec{F}^{\text{grav}}$ on the dust particle due to the Sun just balance the radiation force $\vec{F}^{\text{rad}}$ on it from the sunlight?

SOLUTION ■ We can assume that the Sun is far enough from the particle to act as an isotropic point source of light. Then because we are told that the radiation pressure pushes the particle radially outward from the Sun, we know that the radiation force $\vec{F}^{\text{rad}}$ on the particle is directed radially outward from the center of the Sun. At the same time, the gravitational force $\vec{F}^{\text{grav}}$ on the particle is directed radially inward toward the center of the Sun. Thus, we can write the balance of these two forces as

$$|\vec{F}^{\text{rad}}| = |\vec{F}^{\text{grav}}|. \tag{34-35}$$

Let us consider these forces separately.

Radiation force: To evaluate the left side of Eq. 34-35, we use these three **Key Ideas.**

1. Because the particle is totally absorbing, the force magnitude $|\vec{F}^{\text{rad}}|$ can be found from the intensity I of sunlight at the particle's location and the particle's cross-sectional area A, via Eq. 34-31 ($|\vec{F}| = IA/c$).

2. Because we assume that the Sun is an isotropic point source of light, we can use Eq. 34-25 ($I = P_{\text{sun}}/4\pi r^2$) to relate the Sun's power P_{sun} to the intensity I of the sunlight at the particle's distance r from the Sun.

3. Because the particle is spherical, its cross-sectional area A is πR^2 (*not* half its surface area).

Putting these three ideas together gives us

$$|\vec{F}^{\text{rad}}| = \frac{IA}{c} = \frac{P_{\text{sun}} \pi R^2}{4\pi r^2 c} = \frac{P_{\text{sun}} R^2}{4 r^2 c}. \tag{34-36}$$

Gravitational force: The **Key Idea** here is Newton's law of gravitation (Eq. 14-2), which gives us the magnitude of the gravitational force on the particle as

$$|\vec{F}^{\text{grav}}| = \frac{GM_{\text{sun}} m}{r^2}, \tag{34-37}$$

where M_{sun} is the Sun's mass and m is the particle's mass. Next, the particle's mass is related to its density ρ and volume $V (= \frac{4}{3}\pi R^3$, for a sphere) by

$$\rho = \frac{m}{V} = \frac{m}{\frac{4}{3}\pi R^3}.$$

Solving this for m and substituting the result into Eq. 34-37 give us

$$|\vec{F}^{\text{grav}}| = \frac{GM_{\text{sun}} \rho \left(\frac{4}{3}\pi R^3\right)}{r^2}. \tag{34-38}$$

Then substituting Eqs. 34-36 and 34-38 into Eq. 34-35 and solving for R yield

$$R = \frac{3P_{\text{sun}}}{16\pi c\rho GM_{\text{sun}}}.$$

Using the given value of ρ and the known values of G (Appendix B) and M_{sun} (Appendix C), we can evaluate the denominator:

$$(16\pi)(3 \times 10^8 \text{ m/s})(3.5 \times 10^3 \text{ kg/m}^3)$$
$$\times (6.67 \times 10^{-11} \text{ N} \cdot \text{m}^2/\text{kg}^2)(1.99 \times 10^{30} \text{ kg})$$
$$= 7.0 \times 10^{33} \text{ N/s}.$$

Using P_{sun} from Appendix C, we then have

$$R = \frac{(3)(3.9 \times 10^{26} \text{ W})}{7.0 \times 10^{33} \text{ N/s}} = 1.7 \times 10^{-7} \text{ m}. \quad \text{(Answer)}$$

Note that this result is independent of the particle's distance r from the Sun.

Dust particles with radius $R \approx 1.7 \times 10^{-7}$ m follow an approximately straight path like path b in Fig. 34-10. For larger values of R, comparison of Eqs. 34-36 and 34-38 shows that, because $\vec{F}^{\text{grav}}$ varies with R^3 and $\vec{F}^{\text{rad}}$ varies with R^2, the gravitational force $\vec{F}^{\text{grav}}$ dominates the radiation force $\vec{F}^{\text{rad}}$. Thus, such particles follow a path that is curved toward the Sun like path c in Fig. 34-10. Similarly, for smaller values of R, the radiation force dominates, and the dust follows a path that is curved away from the Sun like path a. The composite of these dust particles is the dust tail of the comet.

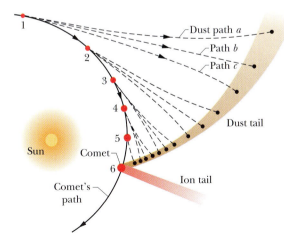

FIGURE 34-10 ■ A comet is now at position 6. Dust it has released at five previous positions has been pushed outward by radiation pressure from sunlight, has taken the dashed paths, and now forms the comet's curved dust tail. Its ion trail points directly away from the Sun.

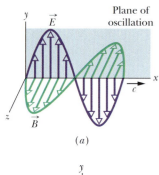

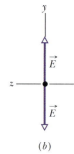

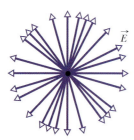

FIGURE 34-11 ■ (a) The plane of oscillation of a polarized electromagnetic wave. (b) To represent the polarization, we view the plane of oscillation "head-on" and indicate the possible directions of the oscillating electric field with two arrows, which we refer to as a "double arrow."

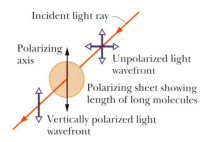

FIGURE 34-12 ■ Unpolarized light consists of waves with randomly directed electric fields. Here the waves are all traveling along the same axis, directly out of the page, and all have the same amplitude $|\vec{E}|$.

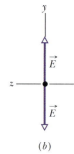

FIGURE 34-13 ■ Unpolarized light becomes polarized when it is sent through a polarizing sheet. The polarization is then parallel to the polarizing axis of the sheet.

34-7 Polarization

At the beginning of the chapter we talked about a charge oscillating along a line. We observed that the charge's oscillation produced kinks in its field lines that are interpreted as wavefronts. We noted that the electric field in the kink was perpendicular to the direction of the propagation and pointed along the line of oscillation of the charge. This means that the electric field in the outgoing wave points back and forth in a single direction. It doesn't wander around pointing in all directions as it would if the electromagnetic wave was the result of many charges oscillating in many different directions. We call such a wave **polarized.** This is an important concept since many of our sources of radiation—such as antennas—impart this property.

Figure 34-11a shows an electromagnetic wave with its electric field oscillating parallel to the vertical y axis. The plane containing the $\vec{E}$ vectors is called the **plane of oscillation** of the wave (hence, the wave is said to be *plane-polarized* parallel to the y axis). We can represent the wave's polarization (state of being polarized) by showing the direction of the electric field oscillations in a "head-on" view of the plane of oscillation, as in Fig. 34-11b. The two vertical arrows in that figure indicate that as the wave travels past us, its electric field oscillates vertically, continuously changing between being directed up and down along the y axis. VHF (very high frequency) television antennas in England are oriented vertically, but those in North America are horizontal. The difference is due to the direction of oscillation of the electromagnetic waves carrying the TV signal. In England, the transmitting equipment is designed to produce waves that are polarized vertically; that is, their electric field oscillates vertically. Thus, for the electric field of the incident television waves to drive a current along an antenna (and provide a signal to a television set), the antenna must be vertical. In North America, the antenna must be horizontal.

Polarized Light

The electromagnetic waves emitted by a television station all have the same polarization because they are generated by electrons moving up and down (or right and left) along a transmission antenna. On the other hand, electromagnetic waves emitted by any common source of *light* (such as the Sun or a lightbulb) are generated by the individual atoms or molecules that comprise the light source. The fact that light is generated by individual atoms or molecules within an object means that there is no preferred *orientation* associated with the electromagnetic waves that make up the light emerging from a source, even in cases where there is a preferred direction of travel.

We call electromagnetic waves with random orientations (like the light produced by atoms and molecules in the sun) **randomly polarized** or **unpolarized.** This is because the direction of the electric field at a given point in space changes direction quickly and randomly. It is still perpendicular to both the direction of travel of the wave and the magnetic field vector on a moment by moment basis, but its orientation in space changes continuously. Figure 34-12 shows an unpolarized electromagnetic wave traveling into or out of the page. If we try to represent a head-on view of the oscillations over some time period, we do not have a simple drawing with a single double arrow like that of Fig. 34-11b; instead we have a mess of double arrows like that in Fig. 34-12.

We can (and often do) transform randomly polarized (or unpolarized) visible light into polarized light by sending it through a *polarizing sheet,* as is shown in Fig. 34-13. Such sheets, commercially known as Polaroids or Polaroid filters, were invented in 1932 by Edwin Land while he was an undergraduate student. A polarizing sheet consists of certain long molecules embedded in plastic. When the sheet is manufactured, it is stretched to align the long molecules in parallel rows, like rows in a plowed field. When light is then sent through the sheet, electric field components perpendicular to the long molecules pass through the sheet, while components parallel to the long molecules are absorbed and disappear.

This is not surprising. The electrons surrounding long molecules are more free to move up and down along the molecular axis and absorb the radiation. Those perpendicular to the long axis are not as free to oscillate. We shall not dwell on the orientation of the molecules but, instead, shall assign a *polarizing axis* to the sheet, along which electric field components are passed:

> An electric field component parallel to the polarizing axis is passed (*transmitted*) by a polarizing sheet; a component perpendicular to it is absorbed.

Thus, the electric field of the light emerging from the sheet consists of only the components that are parallel to the polarizing axis of the sheet. Hence, the light is polarized in that direction. In Fig. 34-13, the vertical electric field components are transmitted by the sheet; the horizontal components are absorbed. The transmitted waves are then vertically polarized.

In some situations, light is **partially polarized** (its field oscillations are not completely random as in Fig. 34-12 nor are they parallel to a single axis as in Fig. 34-11*b*). Partially polarized light can be viewed as a superposition of plane polarized and unpolarized light waves.

Intensity of Transmitted Polarized Light

We now consider the intensity of light transmitted by a polarizing sheet. We start with unpolarized light, whose electric field oscillations we can resolve into y- and z-components as represented in Fig. 34-12*b*. Further, we can arrange for the y axis to be parallel to the polarizing direction of the sheet. Then only the y-components of the light's electric field are passed by the sheet; the z-components are absorbed. As suggested by Fig. 34-12*b*, if the original waves are randomly oriented, the sum of the y-components and the sum of the z-components are equal. When the z-components are absorbed, half the original intensity I_0 of the light is lost. The intensity I of the emerging polarized light is then

$$I = \tfrac{1}{2}I_0. \tag{34-39}$$

Let us call this the *one-half rule;* we can use it *only* when the light reaching a polarizing sheet is unpolarized.

Suppose now that the light reaching a polarizing sheet is already polarized. Figure 34-14 shows a polarizing sheet in the plane of the page and the electric field $\vec{E}$ of such a polarized light wave traveling toward the sheet (and thus prior to any absorption). We can resolve $\vec{E}$ into two components relative to the polarizing axis of the sheet: parallel component E_y is transmitted by the sheet, and perpendicular component E_z is absorbed. Since θ is the angle between $\vec{E}$ and the polarizing axis of the sheet, the transmitted parallel component is

$$E_y = |\vec{E}|\cos\theta. \tag{34-40}$$

Recall that the intensity of an electromagnetic wave (such as our light wave) is proportional to the square of the electric field's magnitude $I = (E^{\text{rms}})^2/c\mu_0$ (Eq. 34-23). In our present case then, the intensity I of the emerging wave is proportional to E_y^2 and

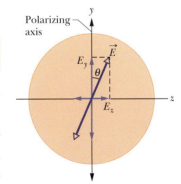

FIGURE 34-14 ■ Polarized light approaching a polarizing sheet. The electric field $\vec{E}$ of the light can be resolved into components E_y (parallel to the polarizing axis of the sheet) and E_z (perpendicular to that axis). Component E_y will be transmitted by the sheet; component E_z will be absorbed. *Note:* The long molecules are oriented perpendicular to the polarizing axis.

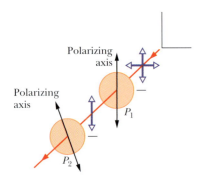

FIGURE 34-15 ■ The light transmitted by polarizing sheet P_1 is vertically polarized, as represented by the vertical double arrow. The amount of that light that is then transmitted by polarizing sheet P_2 depends on the angle between the polarization axis of that light and the polarizing axis of P_2.

the intensity I_0 of the original wave is proportional to $\vec{E}^2$. Hence, from $E_y = |\vec{E}|\cos\theta$ (Eq. 34-40) we can write $I/I_0 = \cos^2\theta$, or

$$I = I_0 \cos^2\theta \qquad \text{(Malus' law).} \tag{34-41}$$

This expression was first introduced in the nineteenth century by Ètienne Malus, a French mathematician who studied polarized light. So, Eq. 34-41 is called Malus' law or the *cosine-squared rule*. We can use it *only* when the light reaching a polarizing sheet is already polarized. Then the transmitted intensity I is a maximum and is equal to the original intensity I_0 when the original wave is polarized parallel to the polarizing axis of the sheet (when θ in Eq. 34-41 is 0° or 180°). I is zero when the original wave is polarized perpendicular to the polarizing axis of the sheet (when θ is 90°).

Figure 34-15 shows an arrangement in which initially unpolarized light is sent through two polarizing sheets P_1 and P_2. (Often, the first sheet is called the *polarizer,* and the second the *analyzer.*) Because the polarizing axis of P_1 is vertical, the light transmitted by P_1 to P_2 is polarized vertically. If the polarizing axis of P_2 is also vertical, then all the light transmitted by P_1 is transmitted by P_2. If the polarizing axis of P_2 is horizontal, none of the light transmitted by P_1 is transmitted by P_2. We reach the same conclusions by considering only the *relative* orientations of the two sheets: If their polarizing axes are parallel, all the light passed by the first sheet is passed by the second sheet. If those axes are perpendicular (the sheets are said to be *crossed*), no light is passed by the second sheet. These two extremes are displayed with polarized sunglasses in Fig. 34-16.

Finally, if the two polarizing axes of Fig. 34-15 make an angle between 0° and 90°, some of the light transmitted by P_1 will be transmitted by P_2. The intensity of that light is determined by $I = I_0 \cos^2\theta$ (Eq. 34-41).

Light can be polarized by means other than polarizing sheets, such as by reflection (discussed in Section 34-10) and by scattering from atoms or molecules. In *scattering,* light that is intercepted by an object, such as a molecule, is sent off in many, perhaps random, directions. An example is the scattering of sunlight by molecules in the atmosphere, which gives the sky its general glow.

Although direct sunlight is unpolarized, light from much of the sky is at least partially polarized by such scattering. Bees use the polarization of sky light in navigating to and from their hives. Similarly, the Vikings used it to navigate across the North Sea when the daytime Sun was below the horizon (because of the high latitude of the North Sea). These early seafarers had discovered certain crystals (now called cordierite) that changed color when rotated in polarized light. By looking at the sky through such a crystal while rotating it about their line of sight, they could locate the hidden Sun and thus determine which way was south.

FIGURE 34-16 ■ Polarizing sunglasses consist of sheets whose polarizing axes are vertical when the sunglasses are worn. (*a*) Overlapping sunglasses transmit light fairly well when their polarizing axes have the same orientation, but (*b*) they block most of the light when they are crossed.

TOUCHSTONE EXAMPLE 34-3: Polarizing Sheets

Figure 34-17 shows a system of three polarizing sheets in the path of initially unpolarized light. The polarizing axis of the first sheet is parallel to the y axis, that of the second sheet is 60° counterclockwise from the y axis, and that of the third sheet is parallel to the x axis. What fraction of the initial intensity I_0 of the light emerges from the system, and how is that light polarized?

SOLUTION ■ The **Key Ideas** here are these:

1. We work through the system sheet by sheet, from the first one encountered by the light to the last one.

2. To find the intensity transmitted by any sheet, we apply either the one-half rule or the cosine-squared rule, depending on whether the light reaching the sheet is unpolarized or already polarized.

3. The light that is transmitted by a polarizing sheet is always polarized parallel to the polarizing axis of the sheet.

First sheet: The original light wave is represented in Fig. 34-17b, using the head-on, double-arrow representation of Fig. 34-11b. Be-

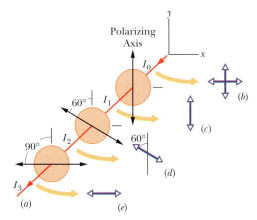

FIGURE 34-17 ■ (a) Initially unpolarized light of intensity I_0 is sent into a system of three polarizing sheets. The intensities I_1, I_2, I_3 of the light transmitted by the sheets are labeled. Shown also are the polzarizations, from head-on views, of (b) the initial light and the light transmitted by (c) the first sheet, (d) the second sheet, and (e) the third sheet.

cause the light is initially unpolarized, the intensity I_1 of the light transmitted by the first sheet is given by the one-half rule (Eq. 34-39):

$$I_1 = \tfrac{1}{2}I_0.$$

Because the polarizing axis of the first sheet is parallel to the y axis, the polarization of the light transmitted by it is also, as shown in the head-on view of Fig. 34-17c.

Second sheet: Since the light reaching the second sheet is polarized, the intensity I_2 of the light transmitted by that sheet is given by the cosine-squared rule (Eq. 34-41). The angle θ in the rule is the angle between the polarization axis of the entering light (parallel to the y axis) and the polarizing axis of the second sheet (60° counterclockwise from the y axis), and so θ is 60°. Then

$$I_2 = I_1 \cos^2 60°.$$

The polarization of this transmitted light is parallel to the polarizing axis of the sheet transmitting it—that is, 60° counterclockwise from the y axis, as shown in the head-on view of Fig. 34-17d.

Third sheet: Because the light reaching the third sheet is polarized, the intensity I_3 of the light transmitted by that sheet is given by the cosine-squared rule. The angle θ is now the angle between the polarization axis of the entering light (Fig. 34-17d) and the polarizing axis of the third sheet (parallel to the x axis), and so $\theta = 30°$. Thus,

$$I_3 = I_2 \cos^2 30°.$$

This final transmitted light is polarized parallel to the x axis (Fig. 34-17e). We find its intensity by substituting first for I_2 and then for I_1 in the equation above:

$$I_3 = I_2 \cos^2 30° = (I_1 \cos^2 60°)\cos^2 30°$$
$$= (\tfrac{1}{2}I_0)\cos^2 60° \cos^2 30° = 0.094 I_0.$$

Thus, $\dfrac{I_3}{I_0} = 0.094.$ (Answer)

That is to say, 9.4% of the initial intensity emerges from the three-sheet system.

If we now remove the second sheet, what fraction of the initial intensity emerges from the system?

34-8 Maxwell's Rainbow

In Maxwell's time (the mid-1800s), the visible, infrared, and ultraviolet forms of light were the only electromagnetic waves known. Spurred on by Maxwell's work, however, Heinrich Hertz discovered what we now call radio waves and verified that they move through the laboratory at the same speed as visible light.

As Fig. 34-18 shows, we now know a wide *spectrum* (or range) of electromagnetic waves, referred to by one imaginative writer as "Maxwell's rainbow" but generally referred to as "light" or "electromagnetic radiation" by physicists.

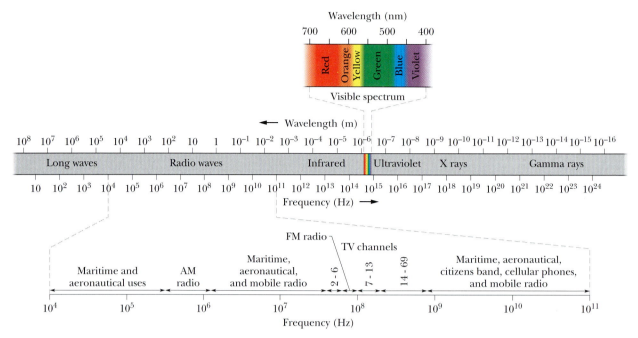

FIGURE 34-18 ■ The electromagnetic spectrum.

Consider the extent to which we are bathed in electromagnetic waves throughout this spectrum. The Sun, whose radiations define the environment in which we as a species have evolved and adapted, is the dominant source. We are also crisscrossed by radio, television and cellular phone signals. Microwaves from radar systems and from telephone relay systems may reach us. There are electromagnetic waves from lightbulbs, from the heated engine blocks of automobiles, from x-ray machines, from lightning flashes, and from buried radioactive materials. Beyond this, radiation reaches us from stars and other objects in our galaxy and from other galaxies. Electromagnetic waves also travel in the other direction. Television signals, transmitted from the Earth since about 1950, have now transmitted news about us (along with episodes of *I Love Lucy*) to whatever technically sophisticated inhabitants there may be on the planets that encircle the nearest 400 or so stars.

In the wavelength scale in Fig. 34-18 (and similarly the corresponding frequency scale), each scale marker represents a change in wavelength (and correspondingly in frequency) by a factor of 10. The scale is open-ended; the wavelengths of electromagnetic waves have no inherent upper or lower bounds.

There are many regions of the electromagnetic spectrum in Fig. 34-18, including radio waves, infrared light produced by the thermal motion of charged particles, visible and ultraviolet light emitted by energetic atoms, x rays that are generated when charged particles collide with solid matter, and gamma rays that originate inside atomic nuclei. These regions denote roughly defined wavelength ranges within which certain kinds of sources and detectors of electromagnetic waves are in common use. Other regions of Fig. 34-18, such as those labeled television and AM radio, represent specific wavelength bands assigned by law for certain commercial or other purposes. There are no gaps in the electromagnetic spectrum—and all electromagnetic waves, no matter where they lie in the spectrum, travel through *free space* (vacuum) with the same speed c.

The visible region of the spectrum is of course of particular interest to us. Figure 34-19 shows the relative sensitivity of the human eye to visible light of various wavelengths. The center of the visible region is about 555 nm, which produces the sensation that we call yellow-green.

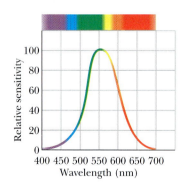

FIGURE 34-19 ■ The relative sensitivity of the average human eye to electromagnetic waves at different wavelengths. This portion of the electromagnetic spectrum to which the eye is sensitive is called visible light.

The limits of this visible spectrum are not well defined because the eye sensitivity curve approaches the zero-sensitivity line asymptotically at both long and short wavelengths. If we take the limits, arbitrarily, as the wavelengths at which eye sensitivity has dropped to 1% of its maximum value, these limits are about 430 and 690 nm. However, the eye can detect electromagnetic waves somewhat beyond these limits if they are intense enough.

Problems

SEC. 34-3 ■ THE GENERATION OF ELECTROMAGNETIC WAVES

1. What Inductance What inductance must be connected to a 17 pF capacitor in an oscillator capable of generating 550 nm (i.e., visible) electromagnetic waves? Comment on your answer.

2. Wavelength What is the wavelength of the electromagnetic wave emitted by the oscillator–antenna system of Fig. 34-4 if $L = 0.253 \ \mu$H and $C = 25.0$ pF?

SEC. 34-4 ■ DESCRIBING ELECTROMAGNETIC WAVE PROPERTIES MATHEMATICALLY

3. Electric Field The electric field of a certain plane electromagnetic wave is given by $E_x = 0$; $E_y = 0$; $E_z = (2.0$ V/m) $\cos[(\pi \times 10^{15} \ \text{s}^{-1})(t - x/c)]$, with $c = 3.0 \times 10^8$ m/s. The wave is propagating in the positive x direction. Write expressions for the components of the magnetic field of the wave.

4. Plane Wave A plane electromagnetic wave has a maximum electric field of 3.20×10^{-4} V/m. Find the maximum magnetic field.

SEC. 34-5 ■ TRANSPORTING ENERGY WITH ELECTROMAGNETIC WAVES

5. Neodymium–Glass Lasers Some neodymium–glass lasers can provide 100 terawatts of power in 1.0 ns pulses at a wavelength of 0.26 μm. How much energy is contained in a single pulse?

6. Poynting Vector Show, by finding the direction of the Poynting vector $\vec{S}$, that the directions of the electric and magnetic fields at all points in Figs. 34-6 to 34-8 are consistent at all times with the assumed directions of propagation.

7. Radiation Emitted The radiation emitted by a laser spreads out in the form of a narrow cone with circular cross section. The angle θ of the cone (see Fig. 34-20) is called the *full-angle beam divergence*. An argon laser, radiating at 514.5 nm,

FIGURE 34-20 ■ Problem 7.

is aimed at the Moon in a ranging experiment. If the beam has a full-angle beam divergence of 0.880 μrad, what area on the Moon's surface is illuminated by the laser?

8. Closest Neighbor Our closest stellar neighbor, Proxima Centauri, is 4.3 lightyears away. It has been suggested that TV programs from our planet have reached this star and may have been viewed by the hypothetical inhabitants of a hypothetical planet orbiting it. Suppose a television station on Earth has a power of 1.0 MW. What is the intensity of its signal at Proxima Centauri?

9. Plane Radio Wave In a plane radio wave the maximum value of the electric field component is 5.00 V/m. Calculate (a) the maximum value of the magnetic field component and (b) the wave intensity.

10. Intensity What is the intensity of a plane traveling electromagnetic wave if $|\vec{B}^{\ \text{max}}|$ is 1.0×10^{-4} T?

11. Maximum Electric Field The maximum electric field at a distance of 10 m from anisotropic point light source is 2.0 V/m. What are (a) the maximum value of the magnetic field and (b) the average intensity of the light there? (c) What is the power of the source?

12. Sunlight Sunlight just outside the Earth's atmosphere has an intensity of 1.40 kW/m². Calculate $|\vec{E}^{\ \text{max}}|$ and $|\vec{B}^{\ \text{max}}|$ for sunlight there, assuming it to be a plane wave.

13. An Airplane An airplane flying at a distance of 10 km from a radio transmitter receives a signal of intensity 10 μW/m². Calculate (a) the amplitude of the electric field at the airplane due to this signal, (b) the amplitude of the magnetic field at the airplane, and (c) the total power of the transmitter, assuming the transmitter to radiate uniformly in all directions.

14. Frank Drake Frank D. Drake, an investigator in the SETI (Search for Extra-Terrestrial Intelligence) program, once said that the large radio telescope in Arecibo, Puerto Rico, "can detect a signal which lays down on the entire surface of the earth a power of only one picowatt." (a) What is the power that would be received by the Arecibo antenna for such a signal? The antenna diameter is 300 m. (b) What would be the power of a source at the center of our galaxy that could provide such a signal? The galactic center is 2.2×10^4 ly away. Take the source as radiating uniformly in all directions.

15. Isotropic Point Source An isotropic point source emits light at wavelength 500 nm, at the rate of 200 W. A light detector is positioned 400 m from the source. What is the maximum rate $\partial B/\partial t$ at which the magnetic component of the light changes with time at the detector's location?

16. Magnetic Component The magnetic component of an electromagnetic wave in vacuum has an amplitude of 85.8 nT and an angular wave number of 4.00 m^{-1}. What are (a) the frequency of the wave, (b) the rms value of the electric component, and (c) the intensity of the light?

17. Magnetic Component Two The magnetic component of a polarized wave of light is

$$B_x = (4.0 \times 10^{-6} \ \text{T}) \sin[(1.57 \times 10^7 \ \text{m}^{-1})y + \omega t]$$

(a) Parallel to which axis is the light polarized? What are the (b) frequency and (c) intensity of the light?

18. Rms Value of Electric Component An electromagnetic wave with a wavelength of 450 nm travels through vacuum in the negative direction of a y axis with its electric component directed parallel to the x axis. The rms value of the electric component is 5.31×10^{-6} V/m. Write an equation for the magnetic component in the form of Eq. 34-3, but complete with numbers.

19. Direct Solar Radiation The intensity of direct solar radiation that is not absorbed by the atmosphere on a particular summer day is 100 W/m². How close would you have to stand to a 1.0 kW electric heater to feel the same intensity? Assume that the heater radiates uniformly in all directions.

20. Isotropic Point Source Two The intensity I of light from an isotropic point light source is determined as a function of the distance r from the source. Figure 34-21 gives intensity I versus the inverse square r^{-2} of that distance. What is the power of the source?

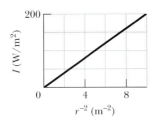

FIGURE 34-21 ▪ Problem 20.

21. What Is the Power During a test, a NATO surveillance radar system, operating at 12 GHz and 180 kW of power, attempts to detect an incoming stealth aircraft at 90 km. Assume that the radar beam is emitted uniformly over a hemisphere. (a) What is the intensity of the beam when it reaches the aircraft's location? The aircraft reflects radar waves as though it has a cross-sectional area of only 0.22 m². (b) What is the power of the aircraft's reflection? Assume that the beam is reflected uniformly over a hemisphere. Back at the radar site, what are the (c) intensity, (d) maximum value of the electric field vector, and (e) rms value of the magnetic field of the reflected (and now detected) radar beam?

22. Average Energy Transport Show that in a plane traveling electromagnetic wave the intensity—that is, the average rate of energy transport per unit area—is given by

$$\langle S \rangle = \frac{(E^{max})^2}{2\mu_0 c} = \frac{(B^{max})^2}{2\mu_0}.$$

SEC. 34-6 ▪ RADIATION PRESSURE

23. High-Power Laser High-power lasers are used to compress a plasma (a gas of charged particles) by radiation pressure. A laser generating pulses of radiation of peak power 1.5 GW is focused onto 1.0 mm² of high-electron-density plasma. Find the pressure exerted on the plasma if the plasma reflects all the light pulses directly back along their paths.

24. Black Cardboard A black, totally absorbing piece of cardboard of area $A = 2.0$ cm² intercepts light with an intensity of 10 W/m² from a camera strobe light. What radiation pressure is produced on the cardboard by the light?

25. Radiation Pressure What is the radiation pressure 1.5 m away from a 500 W lightbulb? Assume that the surface on which the pressure is exerted faces the bulb and is perfectly absorbing and that the bulb radiates uniformly in all directions.

26. Radiation from the Sun Radiation from the Sun reaching the Earth (just outside the atmosphere) has an intensity of 1.4 kW/m². (a) Assuming that the Earth (and its atmosphere) behaves like a flat disk perpendicular to the Sun's rays and that all the incident energy is absorbed, calculate the force on the Earth due to radiation pressure. (b) Compare it with the force due to the Sun's gravitational attraction.

27. Electromagnetic Wave A plane electromagnetic wave, with wavelength 3.0 m, travels in vacuum in the positive x direction with its electric field $\vec{E}$, of amplitude 300 V/m, directed along the y axis. (a) What is the frequency f of the wave? (b) What are the direction and amplitude of the magnetic field associated with the wave? (c) What are the values of k and ω if $\vec{E} = \vec{E}^{max} \sin(kx - \omega t)$? (d) What is the time-averaged rate of energy flow in watts per square meter associated with this wave? (e) If the wave falls on a perfectly absorbing sheet of area 2.0 m², at what rate is momentum delivered to the sheet and what is the radiation pressure exerted on the sheet?

28. Helium–Neon Laser A helium–neon laser of the type often found in physics laboratories has a beam power of 5.00 mW at a wavelength of 633 nm. The beam is focused by a lens to a circular spot whose effective diameter may be taken to be equal to 2.00 wavelengths. Calculate (a) the intensity of the focused beam, (b) the radiation pressure exerted on a tiny perfectly absorbing sphere whose diameter is that of the focal spot, (c) the force exerted on this sphere, and (d) the magnitude of the acceleration imparted to it. Assume a sphere density of 5.00×10^3 kg/m³.

29. Normally Incident Prove, for a plane electromagnetic wave that is normally incident on a plane surface, that the radiation pressure on the surface is equal to the energy density in the incident beam. (This relation between pressure and energy density holds no matter what fraction of the incident energy is reflected.)

30. Laser Beam In Fig. 34-22, a laser beam of power 4.60 W and diameter 2.60 mm is directed upward at one circular face (of diameter $d < 2.60$ mm) of a perfectly reflecting cylinder, which is made to "hover" by the beam's radiation pressure. The cylinder's density is 1.20 g/cm³. What is the cylinder's height H?

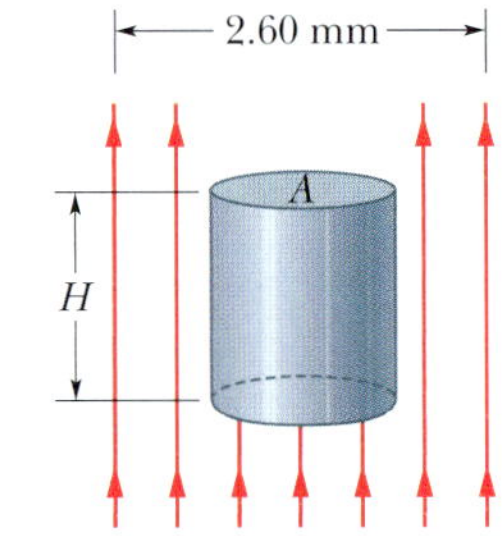

FIGURE 34-22 ▪ Problem 30.

31. Small Spaceship A small spaceship whose mass is 1.5×10^3 kg (including an astronaut) is drifting in outer space with negligible gravitational forces acting on it. If the astronaut turns on a 10 kW laser beam, what speed will the ship attain in 1.0 day because of the momentum carried away by the beam?

32. Average Pressure Prove that the average pressure of a stream of bullets striking a plane surface perpendicularly is twice the kinetic energy density in the stream outside the surface. Assume that the bullets are completely absorbed by the surface. Contrast this with Problem 29.

33. Particle in Solar System A particle in the solar system in under the combined influence of the Sun's gravitational attraction and the radiation force due to the Sun's rays. Assume that the particle is a sphere of density 1.0×10^3 kg/m³ and that all the incident light is absorbed. (a) Show that, if its radius is less than some critical radius R, the particle will be blown out of the solar system. (b) Calculate the critical radius.

34. Radiation Propelled It has been proposed that a spaceship might be propelled in the solar system by radiation pressure, using a large sail made of foil. How large must the sail be if the radiation force is to be equal in magnitude in the Sun's gravitational attraction? Assume that the mass of the ship + sail is 1500 kg, that the sail is perfectly reflecting, and that the sail is oriented perpendicular to the Sun's rays. See Appendix C for needed data. (With a larger sail, the ship is continually driven away from the Sun.)

35. Totally Absorbing Someone plans to float a small, totally absorbing sphere 0.500 m above an isotropic point source of light, so that the upward radiation force from the light matches the downward gravitational force on the sphere. The sphere's density is 19.0 g/cm³ and its radius is 2.00 mm. (a) What power would be required of the light source? (b) Even if such a source were made, why would the support of the sphere be unstable?

36. *frac* Radiation of intensity I is normally incident on an object that absorbs a fraction *frac* of it and reflects the rest back along the original path. What is the radiation pressure on the object?

SEC. 34-7 ■ POLARIZATION

37. Unpolarized Light A beam of unpolarized light of intensity 10 mW/m² is sent through a polarizing sheet as in Fig. 34-13. (a) Find the maximum value of the electric field of the transmitted beam. (b) What radiation pressure is exerted on the polarizing sheet?

38. Magnetic Field Equations The magnetic field equations for an electromagnetic wave in vacuum are $B_x = B \sin(ky + \omega t)$, $B_y = B_z = 0$. (a) What is the direction of propagation? (b) Write the electric field equations. (c) Is the wave polarized? If so, in what direction?

39. Three Polarizing Sheets In Fig. 34-23, initially unpolarized light is sent through three polarizing sheets whose polarizing axes make angles of $\theta_1 = 40°$, $\theta_2 = 20°$, and $\theta_3 = 40°$ with the direction of the y axis. What percentage of the light's initial intensity is transmitted by the system? (*Hint:* Be careful with the angles.)

40. Initially Unpolarized In Fig. 34-23, initially unpolarized light is sent through three polarizing sheets whose polarizing axes make angles of $\theta_1 = \theta_2 = \theta_3 = 50°$ with the direction of the y axis. What percentage of the initial intensity is transmitted by the system of the three sheets? (*Hint:* Be careful with the angles.)

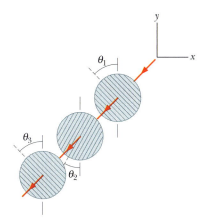

FIGURE 34-23 ■ Problems 39 and 40.

41. Vertically Polarized A horizontal beam of vertically polarized light of intensity 43 W/m² is sent through two polarizing sheets. The polarizing axis of the first is at 70° to the vertical, and that of the second is horizontal. What is the intensity of the light transmitted by the pair of sheets?

42. Two Polarizing Sheets A beam of polarized light is sent through a system of two polarizing sheets. Relative to the polarization axis of that incident light, the polarizing axes of the sheets are at angles θ for the first sheet and 90° for the second sheet. If 0.10 of the incident intensity is transmitted by the two sheets, what is θ?

43. Partially Polarized A beam of partially polarized light can be considered to be a mixture of polarized and unpolarized light. Suppose we send such a beam through a polarizing filter and then

rotate the filter through 360° while keeping it perpendicular to the beam. If the transmitted intensity varies by a factor of 5.0 during the rotation, what fraction of the intensity of the original beam is associated with the beam's polarized light?

44. What Is the Intensity Suppose that in Problem 41 the initial beam is unpolarized. What then is the intensity of the transmitted light?

45. Rotate the Polarization We want to rotate the direction of polarization of a beam of polarized light through 90° by sending the beam through one or more polarizing sheets. (a) What is the minimum number of sheets required? (b) What is the minimum number of sheets required if the transmitted intensity is to be more than 60% of the original intensity?

46. At a Beach At a beach the light is generally partially polarized due to reflections off sand and water. At a particular beach on a particular day near sundown, the horizontal component of the electric field vector is 2.3 times the vertical component. A standing sunbather puts on polarizing sunglasses; the glasses eliminate the horizontal field component. (a) What fraction of the light intensity received before the glasses were put on now reaches the sunbather's eyes? (b) The sunbather, still wearing the glasses, lies on his side. What fraction of the light intensity received before the glasses were put on now reaches his eyes?

47. Four Polarizing Sheets An unpolarized beam of light is sent through a stack of four polarizing sheets, oriented so that the angle between the polarizing directions of adjacent sheets is 30°. What fraction of the incident intensity is transmitted by the system?

48. Four Polarizing Sheets Two In Fig. 34-24, unpolarized light with an intensity of 25 W/m² is sent into a system of four polarizing sheets. What is the intensity of the light that emerges from the system?

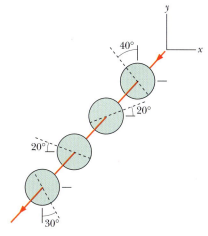

FIGURE 34-24 ■ Problem 48.

49. Two Polarizing Sheets Two A beam of unpolarized light is sent through two polarizing sheets placed one on top of the other. What must be the angle between the polarizing directions of the sheets if the intensity of the transmitted light is to be one-third the incident intensity?

50. Two Polarizing Sheets Three In Fig. 34-25a, unpolarized light is

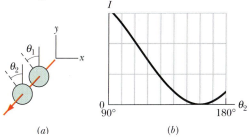

(a)   (b)

FIGURE 34-25 ■ Problem 50.

sent through a system of two polarizing sheets. The angles θ_1 and θ_2 of the polarizing axes of the sheets are measured counterclockwise from the positive direction of the y axis (they are not drawn to scale in the figure). Angle θ_1 is fixed but angle θ_2 can be varied. Figure 34-25b gives the intensity of the light emerging from sheet 2 as a function of θ_2. (The scale of the intensity axis is not indicated.) What percentage of the light's initial intensity is transmitted by the two-sheet system when $\theta_2 = 90°$?

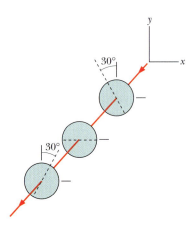
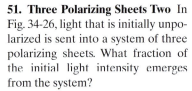

FIGURE 34-26 ■ Problem 51.

51. Three Polarizing Sheets Two In Fig. 34-26, light that is initially unpolarized is sent into a system of three polarizing sheets. What fraction of the initial light intensity emerges from the system?

52. Three Polarizing Sheets Three In Fig. 34-27a, unpolarized light is sent through a system of three polarizing sheets. The angles θ_1, θ_2, and θ_3 of the polarizing axes of the sheets are measured counterclockwise from the positive direction of the y axis (they are not drawn to scale). Angles θ_1 and θ_3 are fixed but angle θ_2 can be varied. Figure 34-28 gives the intensity of the light emerging from sheet 3 as a function of θ_2. (The scale of the intensity axis is not indicated.) What percentage of the light's initial intensity is transmitted by the three-sheet system when $\theta_2 = 90°$?

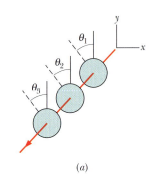

(a)

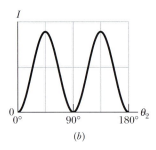

(b)

FIGURE 34-27 ■ Problems 52 and 54.

53. Three Polarizing Sheets Four A system of three polarizing sheets is shown in Fig. 34-29. When initially unpolarized light is sent into the system, the intensity of the transmitted light is 5.0% of the initial intensity. What is the value of θ?

54. Three Polarizing Sheets Five In Fig. 34-27a, unpolarized light is sent through a system of three polarizing sheets. The angles θ_1, θ_2, and θ_3 of the polarizing axes of the sheets are measured counterclockwise from the positive direction of the y axis (they are not drawn to scale). Angles θ_1 and θ_3 are fixed but angle θ_2 can be varied. Figure 34-27b gives the intensity of the light emerging from sheet 3 as a function of θ_2. (The scale of the intensity axis is not indicated.) What percentage of the light's initial intensity is transmitted by the three-sheet system when $\theta_2 = 30°$?

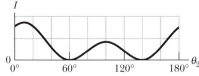

FIGURE 34-28 ■ Problem 52.

SEC. 34-8 ■ MAXWELL'S RAINBOW

55. How Long (a) How long does it take a radio signal to travel 150 km from a transmitter to a receiving antenna? (b) We see a full Moon by reflected sunlight. How much earlier did the light that enters our eye leave the Sun? The Earth–Moon and Earth–Sun distances are 3.8×10^5 km and 1.5×10^8 km. (c) What is the round-trip travel time for light between the Earth and a spaceship orbiting Saturn, 1.3×10^9 km distant? (d) The Crab nebula, which is about 6500 light-years (ly) distant, is thought to be the result of a supernova explosion recorded by Chinese astronomers in A.D. 1054. In approximately what year did the explosion actually occur?

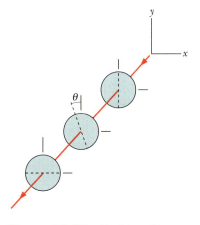

FIGURE 34-29 ■ Problem 53.

56. Project Seafarer Project Seafarer was an ambitious proposal to construct an enormous antenna, buried underground on a site about 10 000 km^2 in area. Its purpose was to transmit signals to submarines while they were deeply submerged. If the effective wavelength were 1.0×10^4 Earth radii, what would be (a) the frequency and (b) the period of the radiations emitted? Ordinarily, electromagnetic radiations do not penetrate very far into conductors such as seawater.

57. At What Wavelengths (a) At what wavelengths does the eye of a standard observer have half its maximum sensitivity? (b) What are the wavelength, frequency, and period of the light for which the eye is the most sensitive?

58. Helium–Neon Laser Two A certain helium–neon laser emits red light in a narrow band of wavelengths centered at 632.8 nm and with a "wavelength width" (such as on the scale of Fig. 34-18) of 0.0100 nm. What is the corresponding "frequency width" for the emission?

59. Speed of Light One method for measuring the speed of light, based on observations by Roemer in 1676, consisted of observing the apparent times of revolution of one of the moons of Jupiter. The true period of revolution is 42.5 h. (a) Taking into account the finite speed of light, how would you expect the apparent time for one revolution to change as the Earth moves in its orbit from point x to point y in Fig. 34-30? (b) What observations would be needed to compute the speed of light? Neglect the motion of Jupiter in its orbit. Figure 34-30 is not drawn to scale.

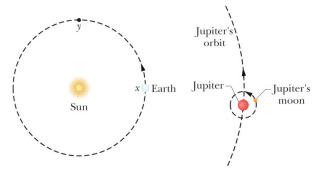

FIGURE 34-30 ■ Problem 59.

Additional Problems

60. Wave of Frequency An electromagnetic wave with frequency 400 terahertz travels through vacuum in the positive direction of an x axis. The wave is polarized, with its electric field directed parallel to the y axis, with amplitude E^{max}. At time $t = 0$, the electric field at point P on the x axis has a value of $+E^{max}/4$ and is decreasing with time. What is the distance along the x axis from point P to the first point with $E = 0$ if we search in (a) the negative direction and (b) the positive direction of the x axis?

61. Earth's Surface At the Earth's surface, what intensity of light is needed to suspend a totally absorbing spherical particle against its own weight if the mass of the particle is 2.0×10^{-13} kg and its radius is 2.0 μm?

62. Tracking a Plane Wave in a Box An oscillating current in an antenna is producing an electromagnetic wave. The region shown in Fig. 34-31 enclosed by a dashed box (not to scale) is far from the antenna. In it, the field produced is well approximated by a plane wave traveling in the z direction and having its E-field pointed along the x direction (using the coordinate system shown).

(a) You perform a series of measurements of the electric field at the origin of your coordinate system and obtain a result that points in the y direction and is well represented by the function

$$E(t) = E_0 \cos(\omega t).$$

What result would you find if instead of at the origin, you repeated the experiment at a point with coordinates $\{0, 0, z\}$? Explain how you know.

(b) What result would you get if you made your measurements at a point in the box with coordinates $\{2, 3, z\}$ cm? (The point is still well within the dashed box.)

(c) For what values of z would you find exactly the same result as you found at the origin?

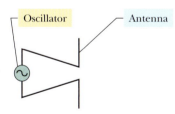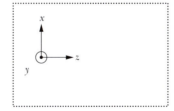

FIGURE 34-31 ■ Problem 62.

63. Electromagnetic Light After completing the construction of his equations for electromagnetism, Maxwell proposed that visible light was actually an electromagnetic wave. Discuss whether or not this is plausible and what evidence there is for his hypothesis.

64. E-Field and String Pulses Compare and contrast the propagation of a pulse on a string and an electromagnetic pulse shown in Fig. 34-32. In particular, address the similarities and differences for
(a) how the pulse "knows" to move from one position to the next;
(b) what will happen if the wave is passed through a slit. (See the figure.)

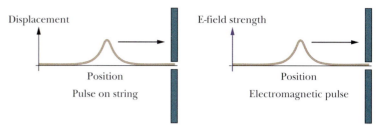

FIGURE 34-32 ■ Problem 64.

65. Measuring the Speed of Light à la Galileo Galileo tried to measure the speed of light by having two people stand on hills about 5 km apart. Each would hold a shuttered lantern. The first would open his lantern and when the second saw the light, he would open his lantern. The first person would then measure how much time it took between the time he first opened his lantern and the time he saw the light returning.

(a) How much time would it take the light to travel between the two hills?

(b) Is this a good way to measure the speed of light? Support your argument with a brief explanation that includes some quantitative discussion of the uncertainty in the measurement.

66. Solar Power for Your House The amount of energy from the sun that reaches the ground is on the order of 1 kW/m². Use this information to estimate the area you would need for a solar energy collector to provide all the electricity in your house. Explain carefully your assumptions and reasoning.

67. Boiling Water in a Microwave Most of you have had the experience of using a microwave oven to boil a cup of water. [If you have not, ask a friend or roommate to help you estimate the time in part (a)]. According to a Pyrex measuring cup that is marked in both English and SI units, one cup contains about 230 ml.

(a) From the amount of time it takes to heat one cup of water from room temperature to boiling in a microwave oven, estimate the power that the oven delivers to the water in watts (joules/second).
(b) Assuming that electromagnetic radiation is flowing into the cup in the microwave from all sides, estimate the electromagnetic energy flux, S, in W/m².
(c) From the flux you calculated in (b), estimate the strength of the electric and magnetic fields in a microwave oven.

68. Speed of Light and the GPS System Although light appears to travel at a speed that is for all practical purposes infinite, for some modern purposes the time delay due to light travel time is of great importance. The Global Positioning System (GPS) allows you to determine your position from comparison of the time delays between radio signals from 4 satellites at a height of 20,000 km above the surface of the earth. (There are actually 24 of these satellites. Your GPS picks out the closest 4 to your current position.) In order to get some idea of how important the speed of light is in establishing your position with one of these gadgets, make some simple assumptions. Assume that a satellite is almost directly overhead. Then figure out how far the satellite will move in the time it takes light (the radio signal) to get from the satellite to your GPS receiver. This estimates how far off the reading of your position would be if your device didn't include the speed of light in its calculations. To do this:

(a) Figure out what speed the satellite must be traveling to be in a circular orbit.

(b) Estimate the time it would take for a radio signal to get from the satellite to your receiver.

(c) Estimate how far the satellite would move in that time. If you ignore light travel time, this tells about how wrong you would get the satellite's position (and therefore how wrong you would get your position).

69. Laser Eye Surgery Laser eye surgery is carried out by delivering highly intense bursts of energy using electromagnetic waves. A typical laser used in such surgery has a wavelength of 190 nm (ultraviolet light) and produces bursts of light that last for 1 ms. The laser delivers an energy of 0.5 mJ to a circular spot on the cornea with a diameter of 1 mm. (The light is well approximated by a plane wave for the short distance between the laser and the cornea.)

(a) Assuming that the energy of a single pulse is delivered to a volume of the cornea about 1 mm³, and assuming that the pulses are delivered so quickly that the energy deposited has no time to

flow out of that volume, how many pulses are required to raise the temperature of that volume from 20°C to 100°C? (Assume that the cornea has a heat capacity similar to that of water.)

(b) Estimate the maximum strength of the electric field in one of these pulses.

70. Insolation of the Earth The power radiated by the sun is 3.9×10^{26} W. The Earth orbits the Sun in a nearly circular orbit of radius 1.5×10^{11} m. The Earth's axis of rotation is tilted by 23° relative to the plane of the orbit (see Fig. 34-33) so sunlight does not strike the equator perpendicularly.

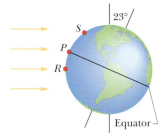

FIGURE 34-33 ■ Problem 70.

(a) At the time of year depicted in Fig. 34-33, what power strikes a 1 m² patch of horizontal flat land at the equator at the point P?

(b) Will a 1 m² patch of horizontal flat land at the point R or S receive more radiation?

(c) Explain how your answer to part (b) tells you at which of the points R or S it is summer or winter.

35 | Images

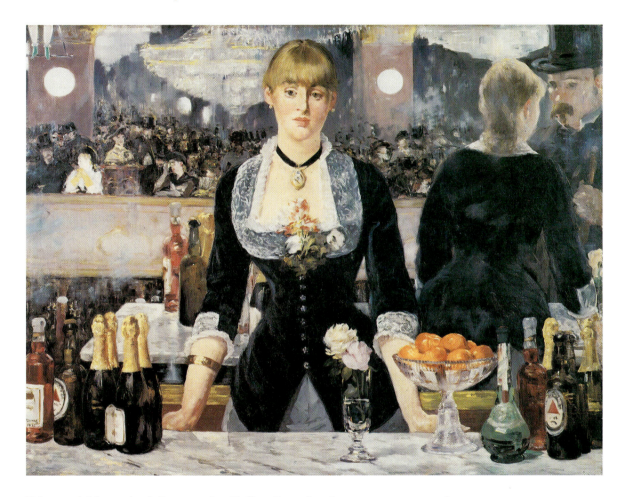

Edouard Manet's *A Bar at the Folies-Bergère* has enchanted viewers ever since it was painted in 1882. Part of its appeal lies in the contrast between an audience ready for entertainment and a bartender whose eyes betray her fatigue. Its appeal also depends on subtle distortions of reality that Manet hid in the painting—distortions that give an eerie feel to the scene even before you recognize what is "wrong."

Can you find those subtle distortions of reality?

The answer is in this chapter.

35-1 Introduction

In Chapter 34 we began our study of electromagnetic waves. In this chapter, our focus is on electromagnetic image formation. Although most of us are familiar with optical images of visible light created by lenses and mirrors, image formation occurs in many other regions of the electromagnetic spectrum, including x rays, ultraviolet and infrared light, microwaves, and radio waves. In this chapter we will concentrate on images formed by visible light.

Electromagnetic waves with wavelengths that are within or near the visually detectable range (typically 400 to 700 nanometers) are commonly referred to as **light.** Visible light and its interaction with materials in our everyday world is of obvious importance. For many species, the ability to **see** (the visual system of an organism's formation of a mental image that is based on the detection of light) is often a necessary condition for survival. Light-based (or **optical**) instruments including eyeglasses, microscopes, and mirrors are important to most people on a daily basis. Whether these instruments have allowed you or someone around you to read the words on a page, diagnose a bacterial infection, or detect the car behind you as you back up in the parking lot, an optical instrument has inevitably impacted your life today. The human eye is an optical instrument.

In Chapter 34, we focused on the wave nature of electromagnetic radiation. In this chapter we will use a simplified wave model call the *ray model* of light. Although a light wave spreads as it moves away from its source, we can often approximate its travel as being in a straight line. For example, we did so in Chapter 34 for the light wave in Fig. 34-7a. This straight-line approximation is the basis of the ray model of light in which light waves are represented as lines called **rays** or **beams.** The study of the properties of light waves under this approximation is called *geometrical optics*. It is a perfectly productive (and simpler) approach to understanding how optical instruments function. We will use it extensively in this chapter.

In order to understand how the optical instruments that are so important in our lives work, we first need to understand some fundamental concepts related to the way light interacts with objects around us. For example,

1. Each small area on the surface of most ordinary objects scatters incident light rays in many directions. So anyone who has a line of sight to parts of an object's surface that are illuminated can see them.

2. Objects with smooth surfaces act as mirrors that reflect a light ray in a single direction rather than scattering it.

3. In transparent materials that are uniform like glass, plastic, air, and water, a ray of light travels in a straight line. However, light rays that cross an interface between two different transparent materials at an angle change direction at that interface.

4. When your eye receives neighboring rays of light that are diverging slightly, your brain assumes that the rays are coming from a common point. This is how your brain constructs a visual model of your surroundings.

In the first few sections of this chapter, we discuss "reflection" and "refraction." These are the two fundamental scattering processes that can occur when light strikes an object with a smooth surface such as a mirror or polished glass. This discussion will provide the foundation required to understand image formation by simple mirrors and lenses, which is covered in Sections 35-6 to 35-10. Finally, we will use our understanding of mirrors and lenses to consider more complex optical devises like microscopes and telescopes. We will continue to use both the wave model and the ray model of light in Chapters 36 and 37.

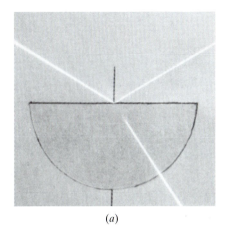

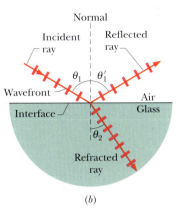

(a)   (b)

FIGURE 35-1 ■ (a) A black-and-white photograph showing the reflection and refraction of an incident beam of light by a horizontal plane glass surface. At the bottom surface, which is curved, the beam is perpendicular to the surface, so the refraction there does not bend the beam. (Note: It is not possible to view a beam of light from the side as it passes through air or a lens. The path of the beams are visible because they are skimming along a piece of paper that is underneath the lens.) (b) A representation of (a) using rays. The angles of incidence (θ_1), of reflection (θ_1'), and of refraction (θ_2) are marked.

35-2 Reflection and Refraction

The black-and-white photograph in Fig. 35-1a shows an example of light waves traveling in approximately straight lines. A narrow beam of incoming light (the **incident beam**), angled downward from the left and traveling through air, encounters a *plane* (flat) glass surface. Part of the light seems to bounce off the smooth glass surface, forming a smooth glass **reflected beam** directed upward toward the right. The rest of the light travels through the surface and into the glass, forming a **refracted beam** directed downward to the right. Since the surface of the glass is smooth and the glass is uniform, the refracted light forms a beam rather than being scattered in many directions. Because light can travel through the glass like this, the glass is said to be *transparent;* that is, we can see through it. (In this chapter we shall consider only transparent materials with smooth surfaces.)

The passage of light from one homogeneous surface to another with a smooth surface (for example, from air to glass) is called **refraction,** and the light is said to be refracted. Unless an incident beam of light is perpendicular to a surface, refraction at the surface (or **interface**) changes the light's direction of travel. For this reason, the beam is said to be "bent" by the refraction. Note that as shown in Fig. 35-1a, the bending occurs only at the surface.

In Figure 35-1b, the beams of light in the photograph are represented with an *incident ray,* a *reflected ray,* and a *refracted ray* (and wave fronts). Each ray is oriented with respect to a line, called the normal, that is perpendicular to the surface at the point of reflection and refraction. In Fig. 35-1b, the **angle of incidence** is θ_1, the **angle of reflection** is θ_1', and the **angle of refraction** is θ_2. These are all measured *relative to the normal* as a line perpendicular to the surface. The plane containing the incident ray and the normal is the plane of incidence, which is in the plane of the page in Fig. 35-1b.

Experiment shows that reflection and refraction from smooth transparent surfaces are governed by two laws:

Law of reflection: A reflected ray lies in the plane of incidence and has an angle of reflection equal to the angle of incidence. In Fig. 35-1b, this means that

$$\theta_1' = \theta_1 \qquad \text{(reflection from a smooth surface).} \qquad (35\text{-}1)$$

(We shall now usually drop the prime on the angle of reflection.) As we will see in the sections that follow, the law of reflection is the fundamental basis for understanding image formation from any kind of mirror. Hence, this simple statement is really quite important.

TABLE 35-1
Some Indices of Refraction[a]

Medium	Index	Medium	Index
Vacuum	Exactly 1	Typical crown glass	1.52
Air (STP)[b]	1.00029	Sodium chloride	1.54
Water (20°C)	1.33	Polystyrene	1.55
Acetone	1.36	Carbon disulfide	1.63
Ethyl alcohol	1.36	Heavy flint glass	1.65
Sugar solution (30%)	1.38	Sapphire	1.77
Fused quartz	1.46	Heaviest flint glass	1.89
Sugar solution (80%)	1.49	Diamond	2.42

[a] For a wavelength of 589 nm (yellow sodium light).
[b] STP means "standard temperature (0°C) and pressure (1 atm)."

Law of refraction (or Snell's law): A refracted ray lies in the plane of incidence and has an angle of refraction θ_2 that is related to the angle of incidence θ_1 by

$$n_2 \sin\theta_2 = n_1 \sin\theta_1 \quad \text{(refraction in a transparent medium).} \quad (35\text{-}2)$$

Here each of the symbols n_1 and n_2 is a dimensionless constant, called the index of refraction, that is associated with a medium involved in the refraction. We derive this equation, called Snell's law, in Chapter 36. As we shall discuss there, the index of refraction, n, of a medium is equal to c/v, where v is the speed of light in that medium and c is its speed in vacuum.

Table 35-1 gives the indices of refraction for visible light of vacuum and some common substances. For vacuum, n is defined to be exactly 1; for air, n is very close to 1.0 (an approximation we shall often make). No material used in basic optical devices has an index of refraction less than 1.

We can rearrange Eq. 35-2 as

$$\sin\theta_2 = \frac{n_1}{n_2} \sin\theta_1 \quad (35\text{-}3)$$

to compare the angle of refraction θ_2 with the angle of incidence θ_1. We can then see that the relative value of θ_2 depends on the relative values of n_2 and n_1. In fact, we can have three basic results:

1. If n_2 is equal to n_1, then θ_2 is equal to θ_1. In this case, refraction does not bend the light beam, which continues in the undeflected direction, as in Fig. 35-2a.

2. If n_2 is greater than n_1, then θ_2 is less than θ_1. In this case, refraction bends the light beam away from the undeflected direction and toward the normal, as in Fig. 35-2b.

FIGURE 35-2 ■ Light refracting from a medium with an index of refraction n_1 and into a medium with an index of refraction n_2. (a) The beam does not bend when $n_2 = n_1$; the refracted light then travels in the *undeflected direction* (the dotted line), which is the same as the direction of the incident beam. The beam bends (b) toward the normal when $n_2 > n_1$ and (c) away from the normal when $n_2 < n_1$.

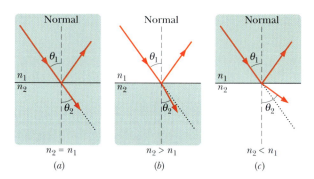

3. If n_2 is less than n_1, then θ_2 is greater than θ_1. In this case, refraction bends the light beam away from the undeflected direction and away from the normal, as in Fig. 35-2c.

Refraction cannot bend a beam so much that the refracted ray is on the same side of the normal as the incident ray.

Chromatic Dispersion

The index of refraction n encountered by light in any medium except a vacuum depends on the wavelength of the light. The dependence of n on wavelength implies that when a light beam consists of rays of different wavelengths, the rays will be refracted at different angles by a surface; that is, the light will be spread out by the refraction. This spreading of light is called **chromatic dispersion,** in which "chromatic" refers to the colors associated with the individual wavelengths (as discussed in Section 34-8) and "dispersion" refers to the spreading of the light according to its wavelengths. The refractions of Figs. 35-1 and 35-2 do not show chromatic dispersion because the beams are *monochromatic* (of a single wavelength or color).

Generally, the index of refraction of a given medium is *greater* for a shorter wavelength (corresponding to, say, blue light) than for a longer wavelength (say, red light). As an example, Fig. 35-3 shows how the index of refraction of fused quartz depends on the wavelength of light. Such dependence means that when a beam with waves of both blue and red light is refracted through a surface, such as from air into quartz or vice versa, the blue *component* (the ray corresponding to the wave of blue light) bends more than the red component.

A beam of *white light* consists of components of all (or nearly all) the colors in the visible spectrum with approximately uniform intensities. When you see such a beam, you perceive white rather than the individual colors. In Fig. 35-4a, a beam of white light in air is incident on a glass surface. (Because the pages of this book are white, a beam of white light is represented with a gray ray here. Also, a beam of monochromatic light is generally represented with a red ray.) Of the refracted light in Fig. 35-4a, only the red and blue components are shown. Because the blue component is bent more than the red component, the angle of refraction θ_{2b} for the blue component is *smaller* than the angle of refraction θ_{2r} for the red component. (Remember, angles are measured relative to the normal.) In Fig. 35-4b, a ray of white light in glass is incident on a glass–air interface. Again, the blue component is bent more than the red component, but now θ_{2b} is greater than θ_{2r}.

To increase the color separation, we can use a solid glass prism with a triangular cross section, as in Fig. 35-5a. The dispersion at the first surface (on the left in Fig. 35-5a, b) is then enhanced by that at the second surface.

The most charming example of chromatic dispersion is a rainbow. As shown in Fig. 35-6, when white sunlight is intercepted by a falling raindrop, some of the light re-

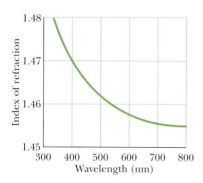

FIGURE 35-3 ■ The index of refraction as a function of wavelength for fused quartz. The graph indicates that a beam of short-wavelength light, for which the index of refraction is higher, is bent more upon entering or leaving quartz than a beam of long-wavelength light.

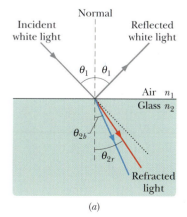

(a)

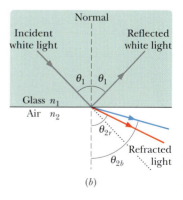

(b)

FIGURE 35-4 ■ Chromatic dispersion of white light. The blue component is bent more than the red component. (a) Passing from air to glass, the blue component ends up with the smaller angle of refraction. (b) Passing from glass to air, the blue component ends up with the greater angle of refraction.

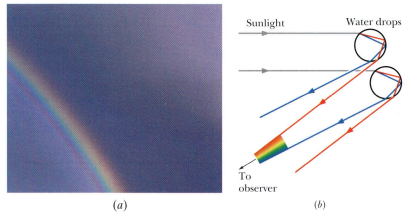

FIGURE 35-5 ■ (*a*) A triangular prism separating white light into its component colors. (*b*) Chromatic dispersion occurs at the first surface and is increased at the second surface.

(*a*)

(*b*)

fracts into the drop and then reflects from the drop's inner surface, via total internal reflection (discussed in the next section). Finally it refracts out of the drop (Fig. 35-6). As with a prism, the first refraction separates the sunlight into its component colors, and the second refraction increases the separation.

The rainbow you see is formed by light that emerges from many drops; the red comes from drops angled slightly higher in the sky, the blue from drops angled slightly lower, and the intermediate colors from drops at intermediate angles. All the drops sending separated colors to you are angled at about 42° from a point that is directly opposite the Sun in your view. If the rainfall is extensive and brightly lit, you see a circular arc of color, with red on top and blue on bottom. Your rainbow is a personal one, because another observer intercepts light from other drops.

FIGURE 35-6 ■ (*a*) A rainbow is always a circular arc that is centered on the direction you would look if you looked directly away from the Sun. Under normal conditions, you are lucky if you see a long arc, but if you are looking downward from an elevated position, you might actually see a full circle. (*b*) The separation of colors when sunlight refracts into and out of falling raindrops leads to a rainbow. The figure shows the situation for the Sun on the horizon (the rays of sunlight are then horizontal). The paths of red and blue rays from two drops are indicated. Many other drops also contribute red and blue rays, as well as the intermediate colors of the visible spectrum.

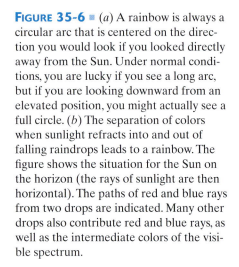

(*a*)

(*b*)

READING EXERCISE 35-1: Which of the three drawings (if any) show physically possible refraction?

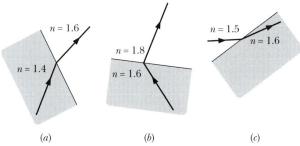

(*a*)   (*b*)   (*c*)

TOUCHSTONE EXAMPLE 35-1:   Reflection and Refraction

(a) In Fig. 35-7a, a beam of monochromatic light reflects and re-fracts at point A on the interface between material 1 with index of refraction $n_1 = 1.33$ and material 2 with index of refraction $n_2 = 1.77$. The incident beam makes an angle of 50° with the interface. What is the angle of reflection at point A? What is the angle of re-fraction there?

SOLUTION ■ The **Key Idea** of any reflection is that the an-gle of reflection is equal to the angle of incidence. Further, both angles are measured between the corresponding light ray and a normal to the interface at the point of reflection. In Fig. 35-7a, the normal at point A is drawn as a dashed line through the point. Note that the angle of incidence θ_1 is not the given 50° but rather is $90° - 50° = 40°$. Thus, the angle of reflection is

$$\theta_1' = \theta_1 = 40°. \qquad \text{(Answer)}$$

The light that passes from material 1 into material 2 undergoes refraction at point A on the interface between the two materials. The **Key Idea** of any refraction is that we can relate the angle of incidence, the angle of refraction, and the indexes of refraction of the two materials via Eq. 35-2:

$$n_2 \sin \theta_2 = n_1 \sin \theta_1. \qquad (35\text{-}4)$$

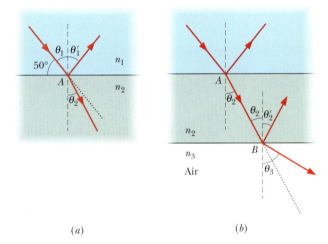

(a)

(b)

FIGURE 35-7 ■ (a) Light reflects and refracts at point A on the interface between materials 1 and 2. (b) The light that passes through material 2 reflects and refracts and point B on the interface between materials 2 and 3 (air).

Again we measure angles between light rays and a normal, here at the point of refraction. Thus, in Fig. 35-7a, the angle of refraction is the angle marked θ_2. Solving Eq. 35-4 for θ_2 gives us

$$\theta_2 = \sin^{-1}\left(\frac{n_1}{n_2} \sin \theta_1\right) = \sin^{-1}\left(\frac{1.33}{1.77} \sin 40°\right)$$

$$= 28.88° \approx 29°. \qquad \text{(Answer)}$$

This result means that the beam swings toward the normal (it was at 40° to the normal and is now at 29°). The reason is that when the light travels across the interface, it moves into a material with a greater index of refraction.

(b) The light that enters material 2 at point A then reaches point B on the interface between material 2 and material 3, which is air, as shown in Fig. 35-7b. The interface through B is parallel to that through A. At B, some of the light reflects and the rest enters the air. What is the angle of reflection? What is the angle of refraction into the air?

SOLUTION ■ We first need to relate one of the angles at point B with a known angle at point A. Because the interface through point B is parallel to that through point A, the incident angle at B must be equal to the angle of refraction θ_2, as shown in Fig. 35-7b. Then for reflection, we use the same **Key Idea** as in (a): the law of reflection. Thus, the angle of reflection at B is

$$\theta_2' = \theta_2 = 28.88° \approx 29°. \qquad \text{(Answer)}$$

Next, the light that passes from material 2 into the air undergoes re-fraction at point B, with refraction angle θ_3. Thus, the **Key Idea** here is again to apply the law of refraction, but this time by writing Eq. 35-4 as

$$n_3 \sin \theta_3 = n_2 \sin \theta_2. \qquad (35\text{-}5)$$

Solving for θ_3 then leads to

$$\theta_3 = \sin^{-1}\left(\frac{n_2}{n_3} \sin \theta_2\right) = \sin^{-1}\left(\frac{1.77}{1.00} \sin 28.88°\right)$$

$$= 58.75° \approx 59°. \qquad \text{(Answer)}$$

This result means that the beam swings away from the normal (it was at 29° to the normal and is now at 59°). The reason is that when the light travels across the interface, it moves into a material (air) with a lower index of refraction.

35-3 Total Internal Reflection

Figure 35-8 shows rays of monochromatic light from a point source S in glass incident on the interface between the glass and air. For ray a, which is perpendicular to the in-terface, part of the light reflects at the interface and the rest travels through it with no change in direction.

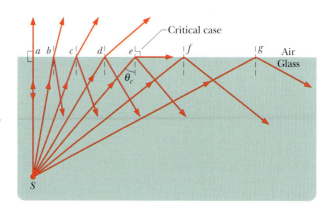

FIGURE 35-8 ■ Total internal reflection of light from a point source S in glass occurs for all angles of incidence greater than the critical angle θ_c. At the critical angle, shown at point e, the refracted ray points along the air–glass interface.

For rays b through e, which have progressively larger angles of incidence at the interface, there are also both reflection and refraction at the interface. As the angle of incidence increases, the angle of refraction increases; for ray e it is 90°, which means that the refracted ray points directly along the interface. The angle of incidence that gives this situation is called the critical angle θ_c. For angles of incidence larger than θ_c, such as for rays f and g, there is no refracted ray and all the light is reflected; this effect is called total internal reflection.

To find θ_c, we use Eq. 35-2; we arbitrarily associate subscript 1 with the glass and subscript 2 with the air, and then we substitute θ_c for θ_1 and 90° for θ_2, finding

$$n_1 \sin \theta_c = n_2 \sin 90°,$$

which gives us

$$\theta_c = \sin^{-1}\frac{n_2}{n_1} \qquad \text{(critical angle)}. \qquad (35\text{-}6)$$

Because the sine of an angle cannot exceed unity, n_2 cannot exceed n_1 in this equation. This restriction tells us that total internal reflection cannot occur when the incident light is in the medium of lower index of refraction. If source S were in the air in Fig. 35-8, all its rays that are incident on the air–glass interface (including f and g) would be both reflected and refracted at the interface.

Total internal reflection has found many applications in medical technology. For example, a physician can search for an ulcer in the stomach of a patient by running two thin bundles of *optical fibers* (Fig. 35-9) down the patient's throat. Light introduced at the outer end of one bundle undergoes repeated total internal reflection within the fibers so that, even though the bundle provides a curved path, most of the light ends up exiting the other end and illuminating the interior of the stomach. Some of the light reflected from the interior then comes back up the second bundle in a similar way, to be detected and converted to an image on a monitor's screen for the physician to view.

FIGURE 35-9 ■ Light sent into one end of an optical fiber like those shown here is transmitted to the opposite end with little loss of light through the sides of the fiber.

TOUCHSTONE EXAMPLE 35-2: Triangular Prism

Figure 35-10 shows a triangular prism of glass in air; an incident ray enters the glass perpendicular to one face and is totally reflected at the far glass–air interface as indicated. If θ_1 is 45°, what can you say about the index of refraction n of the glass?

SOLUTION ■ One **Key Idea** here is that because the light ray is totally reflected at the interface, the critical angle θ_c for that interface must be less than the incident angle of 45°. A second **Key**

Idea is that we can relate the index of refraction n of the glass to θ_c with the law of refraction, Eq. 35-2. Substituting $n_2 = 1$ (for the

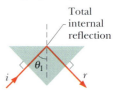

FIGURE 35-10 ■ The incident ray i is totally internally reflected at the glass–air interface, becoming the reflected ray r.

air) and $n_1 = n$ (for the glass) into Eq. 35-2 yields

$$\theta_c = \sin^{-1}\frac{n_2}{n_1} = \sin^{-1}\frac{1}{n}.$$

Because θ_c must be less than the incident angle of 45°, and the sine function is increasing between 0° and 90°,

$$\sin^{-1}\frac{1}{n} < 45°,$$

which gives us

$$\frac{1}{n} < \sin 45°$$

or

$$n > \frac{1}{\sin 45°} = 1.4. \qquad \text{(Answer)}$$

The index of refraction of the glass must be greater than 1.4; otherwise, total internal reflection would not occur for the incident ray shown.

35-4 Polarization by Reflection

As we discuss in Chapter 34, you can increase and decrease the glare you see in sunlight that has been reflected from, say, water by looking through a polarizing sheet (such as a polarizing sunglass lens) and then rotating the sheet's polarizing axis around your line of sight. You can do so because reflected light is fully or partially polarized by the reflection from a surface.

Figure 35-11 shows a ray of unpolarized light incident on a glass surface. Let us resolve the electric field vectors of the light into two components. The *perpendicular components* are perpendicular to the plane of incidence and thus also to the page in Fig. 35-11; these components are represented with dots (as if we see the tips of the vectors). The *parallel components* are parallel to the plane of incidence and the page; they are represented with double-headed arrows. Because the light is unpolarized, these two components are of equal magnitude.

In general, the reflected light also has both components but with unequal magnitudes. This means that the reflected light is partially polarized—the electric fields oscillating along one direction have greater amplitudes than those oscillating along other directions. However, when the light is incident at a particular incident angle, called the *Brewster angle* θ_B, the reflected light has only perpendicular components, as shown in Fig. 35-11. The reflected light is then fully polarized perpendicular to the plane of incidence. The parallel components of the incident light do not disappear; they and perpendicular components form the light that is refracted through the glass surface.

Glass, water, and the other dielectric materials discussed in Section 28-6 can partially and fully polarize light by reflection. When you intercept sunlight reflected from such a surface, you see a bright spot (the glare) on the surface where the reflection takes place. If the surface is horizontal as in Fig. 35-11, the reflected light is partially or fully polarized horizontally. To eliminate such glare from horizontal surfaces, the lenses in polarizing sunglasses are mounted with their polarizing direction vertical.

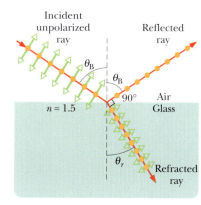

• Component perpendicular to page
↤↦ Component parallel to page

FIGURE 35-11 ■ A ray of unpolarized light in air is incident on a glass surface at the Brewster angle θ_B. The electric fields along that ray have been resolved into components perpendicular to the page (the plane of incidence) and components parallel to the page. The reflected light consists only of components perpendicular to the page and is thus polarized in that direction. The refracted light consists of the original components parallel to the page and weaker components perpendicular to the page; this light is partially polarized.

Brewster's Law

For light incident at the Brewster angle θ_B, we find experimentally that the reflected and refracted rays are perpendicular to each other. Because the reflected ray is reflected at the angle θ_B in Fig. 35-11 and the refracted ray is at an angle $\theta_2 = \theta_r$, we have

$$\theta_B + \theta_r = 90°. \qquad (35-7)$$

These two angles can also be related with Eq. 35-2. Arbitrarily assigning subscript 1 in Eq. 35-2 to the material through which the incident and reflected rays travel, we have, from that equation,

$$n_1 \sin \theta_B = n_2 \sin \theta_r.$$

Combining these equations leads to

$$n_1 \sin \theta_B = n_2 \sin(90° - \theta_B) = n_2 \cos \theta_B,$$

which gives us

$$\theta_B = \tan^{-1} \frac{n_2}{n_1} \qquad \text{(Brewster angle).} \qquad (35\text{-}8)$$

(Note carefully that the subscripts in Eq. 35-8 are not arbitrary because of our decision as to their meaning.) If the incident and reflected rays travel *in air*, we can approximate n_1 as unity and let n represent n_2 in order to write Eq. 35-8 as

$$\theta_B = \tan^{-1} n \qquad \text{(only if } n_1 = 1\text{).} \qquad (35\text{-}9)$$

This simplified version of Eq. 35-8 is also named after Sir David Brewster, who verified both equations experimentally in 1812.

35-5 Two Types of Image

Up to this point, most of our study of physics in general and optics in particular has been focused on understanding physical phenomena. We have not had to pay particular attention to the role of the observer in determining information about the physical system. (Even though there always is such an observer assumed, we also assume the measurements can be made gently enough so as not to disturb the system so the observer can be ignored.) That will need to change now. In studying image formation, we cannot be concerned only with what happens to the light. We must also be concerned with what it looks like to the observer. For this reason, we have to think about how your eyes and brain interpret the signals they receive. This new concern with the observer continues as we move into the study of relativity in Chapter 38 and as you move into quantum physics in your later studies.

For you to see an object, your eye must intercept some of the light rays spreading from the object and then redirect them onto the retina at the rear of the eye. Your visual system, starting with the retina and ending with the visual cortex at the rear of your brain, automatically and subconsciously processes the information provided by the light. That system identifies edges, orientations, textures, shapes, and colors and then rapidly brings to your consciousness an **image** (a reproduction derived from light) of the object; you perceive and recognize the object as being in the direction from which the light rays came and at the proper distance.

Your visual system goes through this processing and recognition even if the light rays do not come directly from the object, but instead reflect toward you from a mirror or refract through the lenses in a pair of binoculars. However, independent of whether the light rays come directly from an object or indirectly from a reflection or refraction event, the visual system in the human brain always forms an image as follows:

> The apparent location of an object is the common point from which the diverging straight line light rays seem to have come (even if the light rays have actually been bent). See Figs. 35-13 and 35-14 for examples.

For example, if the light rays have been reflected toward you from a standard flat mirror, the object appears to be behind the mirror because the rays you intercept come from that direction. Of course, the object is not back there. This type of image, which is called a **virtual image,** truly exists only on your retina but nevertheless is *said* to exist at the perceived location.

A **real image** differs in that it can be formed on a surface, such as a card or a movie screen. You can see a real image (otherwise movie theaters would be empty), but the existence of the image does not depend on your seeing it and it is present even if you are not.

In this chapter we explore several ways in which virtual and real images are formed by reflection (as with mirrors) and refraction (as with lenses). We also distinguish between the two types of image more carefully. We start by considering an example of a natural virtual image.

A Common Mirage

A common example of a virtual image is a pool of water that appears to lie on the road some distance ahead of you on a sunny day, but that you can never reach. The pool is a *mirage* (a type of illusion), formed by light rays coming from the low section of the sky in front of you (Fig. 35-12a). As the rays approach a road that has been heated by the Sun, they travel through progressively warmer air that has been heated by the road. With an increase in air temperature, the speed of light in air increases slightly and, correspondingly, the index of refraction of the air closer to the road decreases continuously. Thus, as the rays descend, encountering progressively smaller indices of refraction, they gradually bend more and more toward the horizontal (Fig. 35-12b).

Once a ray is horizontal, somewhat above the road's surface, it still bends because the lower portion of each associated wave front is in slightly warmer air and is moving slightly faster than the upper portion of the wavefront (Fig. 35-12c). This nonuniform motion of the wavefronts bends the ray upward. As the ray then ascends, it continues to bend upward through progressively greater indexes of refraction (Fig. 35-12d).

If some of this light enters your eyes, your visual system automatically infers that it originated along a backward extension of the rays you have intercepted and, to make sense of the light, assumes that it came from the road surface. If the light happens to be bluish from blue sky, the mirage appears bluish, like water. Because the air is probably turbulent due to the heating, the mirage shimmies, as if water waves were present. The bluish coloring and the shimmy enhance the illusion of a pool of water, but you are actually seeing a virtual image of a low section of the sky.

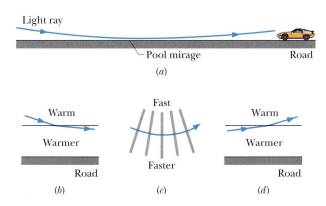

FIGURE 35-12 ◾ (a) A ray from a low section of the sky refracts through air that is heated by a road (without reaching the road). An observer who intercepts the light perceives it to be from a pool of water on the road. (b) Bending (exaggerated) of a light ray descending across an imaginary boundary from warm air to warmer air. (c) Shifting of wavefronts and associated bending of a ray, which occurs because the lower ends of wavefronts move faster in warmer air. (d) Bending of a ray ascending across an imaginary boundary to warm air from warmer air.

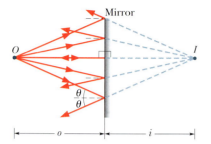

FIGURE 35-13 ■ A point source of light O, called the object, is a perpendicular distance o in front of a plane mirror. Light rays reaching the mirror from O reflect from the mirror. If your eye intercepts some of the reflected rays, you perceive a point source of light I to be behind the mirror, at a perpendicular distance i. The perceived source I is a virtual image of object O.

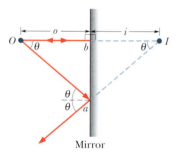

FIGURE 35-14 ■ Two rays from Fig. 35-13. Ray Oa makes an arbitrary angle θ with the normal to the mirror surface. Ray Ob is perpendicular to the mirror.

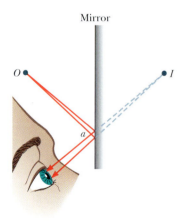

FIGURE 35-15 ■ A "pencil" of rays from O enters the eye after reflection at the mirror. Only a small portion of the mirror near a is involved in this reflection. The light appears to originate at point I behind the mirror.

35-6 Plane Mirrors

A **mirror** is a very smooth surface that can reflect a beam of light (or other electromagnetic radiation) in one direction instead of either scattering it widely in many directions or absorbing it. A shiny metal surface acts as a mirror; a concrete wall does not. In this section we examine the images that a **plane mirror** (a flat reflecting surface) can produce.

Figure 35-13 shows a point source of light O, which we shall call the *object*, at a perpendicular distance o in front of a plane mirror. The light that is incident on the mirror is represented with rays spreading from O.

Caution: Since O is a point while o is a distance, it is important to distinguish between the two, not to confuse them with zero, and to write each one differently.

The reflection of that light is represented with reflected rays spreading from the mirror. If we extend the reflected rays backward (behind the mirror), we find that the extensions intersect at a point that is a perpendicular distance i behind the mirror.

If you look into the mirror of Fig. 35-13, your eyes intercept some of the reflected light. To make sense of what you see, you perceive a point source of light located at the point of intersection of the extensions. This point source is the image I of object O. It is called a point image because it is a point, and it is a virtual image because the rays do not actually pass through it. (As you will see, rays do pass through a point of intersection for a real image.) Your eyes (and brain) trace the rays back to their apparent intersection point and are fooled into thinking that's where the object is.

Figure 35-14 shows two rays selected from the many rays in Fig. 35-13. One reaches the mirror at point b, perpendicularly. The other reaches it at an arbitrary point a, with an angle of incidence θ. The extensions of the two reflected rays are also shown. The right triangles $aOba$ and $aIba$ have a common side and three equal angles and are thus congruent, so their horizontal sides are congruent. That is,

$$Ib = Ob, \qquad (35\text{-}10)$$

where Ib and Ob are the distances from the mirror to the image and the object, respectively. Equation 35-10 tells us that the image is as far behind the mirror as the object is in front of it. By convention (that is, to get our equations to work out), *object distances o* are taken to be positive quantities, and image distances i for virtual images (as here) are taken to be negative quantities. Thus, Eq. 35-10 can be written as $|i| = o$, or as

$$i = -o \qquad \text{(plane mirror)}. \qquad (35\text{-}11)$$

Only rays that are fairly close together can enter the eye after reflection at a mirror. For the eye position shown in Fig. 35-15, only a small portion of the mirror near point a (a portion smaller than the pupil of the eye) is useful in forming the image. To find this portion, close one eye and look at the mirror image of a small object such as the tip of a pencil. Then move your fingertip over the mirror surface until you cannot see the image. Only light coming from that small portion of the mirror under your fingertip produced the image.

Extended Objects

In Fig. 35-16, an extended object O, represented by an upright arrow, is at perpendicular distance o in front of a plane mirror. Each small portion of the object that faces the mirror acts like the point source O of Figs. 35-13 and 35-14. If you intercept the light reflected by the mirror, you perceive a virtual image I that is a composite of the virtual point images of all those portions of the object and seems to be at distance i behind the mirror. Distances i and o are related by Eq. 35-11.

We can also locate the image of an extended object as we did for a point object in Fig. 35-13: we draw some of the rays that reach the mirror from the top of the object, draw the corresponding reflected rays, and then extend those reflected rays behind the mirror until they intersect to form an image of the top of the object. We then do the same for rays from the bottom of the object. As shown in Fig. 35-16, we find that virtual image I has the same orientation and *height* (measured parallel to the mirror) as object O.

Manet's *A Bar at the Folies-Bergère*

In *A Bar at the Folies-Bergère* you see the barroom via reflection by a large mirror on the wall behind the woman tending bar, but the reflection is subtly wrong in three ways. First note the bottles at the left. Manet painted their reflections in the mirror but misplaced them, painting them farther toward the front of the bar than they should be.

Now note the reflection of the woman. Since your view is from directly in front of the woman, her reflection should be behind her, with only a little of it (if any) visible to you; yet Manet painted her reflection well off to the right. Finally, note the reflection of the man facing her. He must be you, because the reflection shows that he is directly in front of the woman, and thus he must be the viewer of the painting. You are looking into Manet's work and seeing your reflection well off to your right. The effect is eerie because it is not what we expect from either a painting or a mirror.

FIGURE 35-16 ■ An extended object O and its virtual image I in a plane mirror.

READING EXERCISE 35-2: In the figure you look into a system of two vertical parallel mirrors A and B separated by distance d. A grinning gargoyle is perched at point O, a distance $0.2d$ from mirror A. Each mirror produces a *first* (least deep) image of the gargoyle. Then each mirror produces a *second* image with the object being the first image in the opposite mirror. Then each mirror produces a *third* image with the object being the second image in the opposite mirror, and so on — you might see hundreds of grinning gargoyle images. How deep behind mirror A are the first, second, and third images in mirror A?

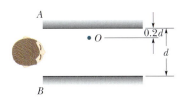

READING EXERCISE 35-3: Is an object in a mirror reversed left to right? How does this happen? Why isn't it also upside down? If you look at your image in a flat mirror, your left hand becomes your right hand, but your head and feet are not interchanged. Can you explain why?

35-7 Spherical Mirrors

We turn now from images produced by plane mirrors to images produced by mirrors with curved surfaces. In particular, we shall consider spherical mirrors, which are simply mirrors in the shape of a small section of the surface of a sphere. A plane mirror is in fact a spherical mirror with an infinitely large *radius of curvature*. This is like treating a small piece of the Earth as if it were flat. If the Earth has a large enough radius, we can't tell the difference.

A plane mirror fooled your visual system into thinking an object was in a different place than it really was by bending the rays so they didn't come directly from the object in straight lines. A curved mirror does even more interesting things by bending the rays in a different way.

Making a Spherical Mirror

We start with the plane mirror of Fig. 35-17*a*, which faces leftward toward an object *O* and an observer. We make a **concave mirror** by curving the mirror's surface so it is concave ("caved in") as in Fig. 35-17*b*. Curving the surface in this way changes several characteristics of the mirror and the image it produces of the object:

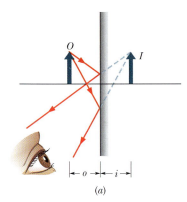

(a)

1. *The center of curvature C* (the center of the sphere of which the mirror's surface is part) was infinitely far from the plane mirror; it is now closer and in front of the concave mirror.

2. *The field of view*—the extent of the scene that is reflected to the observer—was wide for the plane mirror; it is now smaller.

3. The image of the object was as far behind the plane mirror as the object was in front; the image is farther behind the concave mirror; that is, $|i|$ is greater.

4. The height or size of the image was equal to the height or size of the object for the plane mirror. The height of the image is now greater than the height of the object. This feature is why many makeup mirrors and shaving mirrors are concave—they produce a larger image of a face.

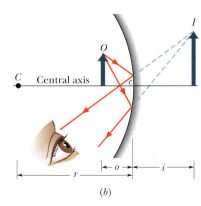

(b)

We can make a **convex mirror** by curving a plane mirror so its surface is *convex* ("flexed out") as in Fig. 35-17*c*. Curving the surface in this way:

1. Moves the center of curvature *C* to *behind* the mirror.

2. *Increases* the field of view. It is wider with a convex mirror than with a plane mirror.

3. Moves the image of the object *closer* to the mirror as compared to the plane mirror.

4. *Shrinks* the size of the image. It is now smaller than the actual size of the object.

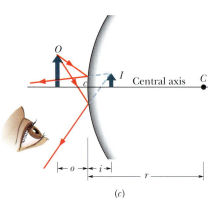

(c)

FIGURE 35-17 ■ (*a*) An object *O* forms a virtual image *I* in a plane mirror. (*b*) If the mirror is bent so that it becomes *concave*, the image moves farther away and becomes larger. (*c*) If the plane mirror is bent so that it becomes *convex*, the image moves closer and becomes smaller.

Side view mirrors on cars and store surveillance mirrors are usually convex to take advantage of the increase in the field of view—more of the store can then be monitored with a single mirror.

In looking at Fig. 35-17*b* and *c*, you should note that when both surfaces of a curved mirror are reflective, the side the observer is on determines whether the mirror is convex or concave.

Focal Points of Spherical Mirrors

In order to figure out how a ray of light reflects from a curved mirror, we will look at a very small region of mirror around the point that the ray we are considering strikes. For the small enough region, the mirror looks flat (like a bit of the Earth looks flat although it is actually curved). Then, we use our plane mirror principle from above: angle of incidence is equal to angle of reflection (Eq. 35-1). To do this, first we need to draw a normal to the surface of the mirror at the point that the ray strikes. Then,

> For a spherical mirror, the reflected ray is in the plane determined by the incident ray and the normal to the surface. The angle between the normal and the reflected ray is equal to the angle between the normal and the incident ray.

For a plane mirror, the magnitude of the image distance i is always equal to the object distance o. Before we can determine how these two distances are related for a spherical mirror, we find it convenient to consider the reflection of light from an object O located an effectively infinite distance in front of a spherical mirror, on the mirror's *central axis*. That axis extends through the center of curvature C and the center c of the mirror. Because of the great distance between the object and the mirror, the light waves spreading from the object are nearly plane waves when they reach the mirror along the central axis. This means that the rays representing the light waves are all parallel to the central axis when they reach the mirror.

When these parallel rays reach a concave mirror like that of Fig. 35-18a, those near the central axis are reflected through a common point F; two of these reflected rays are shown in the figure. If we placed a (small) card at F, a point image of the infinitely distant object O would appear on the card since rays actually converge at that point. (This would occur for any infinitely distant object.) Point F is called the **focal point** (or **focus**) of the mirror, and its distance from the center of the mirror is the **focal length** f of the mirror.

If we now substitute a convex mirror for the concave mirror, we find that the parallel rays are no longer reflected through a common point. Instead, they diverge as shown in Fig. 35-18b. However, if your eye or a camera lens intercepts some of the reflected light, you perceive the light as originating from a point source behind the mirror. Although no rays actually converge behind the convex mirror, this perceived source is located where extensions of the reflected rays pass through a common point (F in Fig. 35-18b). That point is the focal point (or focus) F of the convex mirror, and its distance from the mirror surface is the focal length f of the mirror. If we placed a card at this focal point, an image of object O would not appear on the card, so this focal point is a virtual focal point and is *not* like that of a concave mirror.

To distinguish the actual focal point of a concave mirror from the perceived focal point of a convex mirror, the former is said to be a *real focal point* and the latter is said to be a *virtual focal point*. Moreover, the focal length f of a concave mirror is taken to be a positive quantity, and that of a convex mirror a negative quantity. For mirrors of both types, the focal length f is related to the radius of curvature r of the mirror by

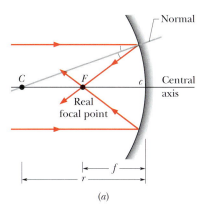

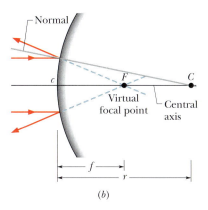

FIGURE 35-18 ■ (a) In a concave mirror, incident parallel light rays are brought to a real focus at F, on the same side of the mirror as the light rays. (b) In a convex mirror, incident parallel light rays seem to diverge from a virtual focus at F, on the side of the mirror opposite the light rays.

$$f = \tfrac{1}{2}r \quad \text{(spherical mirror)}, \qquad (35\text{-}12)$$

where, consistent with the signs for the focal length, r is a positive quantity for a concave mirror and a negative quantity for a convex mirror.

35-8 Images from Spherical Mirrors

With the focal point of a spherical mirror defined, we can find the relation between image distance i and object distance o for concave and convex spherical mirrors. We discussed the law of reflection (Eq. 35-1) in Section 35-2. This law states that the angle of reflection is equal to the angle of incidence. It alone is the foundation required to understand the change in direction of light rays at the surface of a mirror. Ultimately, it is the direction of the reflected light rays that determines where the image is perceived to be located.

We begin by placing the object O *inside the focal point* of the concave mirror—that is, between the mirror and its focal point F (Fig. 35-19a). An observer can then see a virtual image of O in the mirror: The image appears to be behind the mirror, and it has the same orientation as the object.

If we now move the object away from the mirror until it is at the focal point, the image moves farther back from the mirror until it is at infinity (Fig. 35-19b).

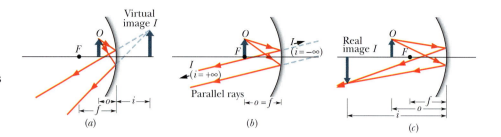

FIGURE 35-19 ■ (a) An object O inside the focal point of a concave mirror, and its virtual image I. (b) The object at the focal point F. (c) The object outside the focal point, and its real image I.

The image is then ambiguous and imperceptible because neither the rays reflected by the mirror nor the ray extensions behind the mirror cross to form an image of O.

If we next move the object *outside the focal point*—that is, farther away from the mirror than the focal point—the rays reflected by the mirror converge to form an *inverted* image of object O (Fig. 35-19c) in front of the mirror. That image moves in from infinity as we move the object farther outside F. If you were to hold a card at the position of the image, the rays converging at that point would scatter in all directions, when those scattered rays entered your eyes, your brain would see them as coming from the card, and so an image of the object would appear on the card. In this case, the image is said to be *focused* on the card by the mirror. (The verb "focus," which in this context means to produce an image, differs from the noun "focus," which is another name for the focal point.) Because this image can actually appear on a surface, it is a real image—the rays actually intersect to create the image, regardless of whether an observer is present. The image distance i of a real image is a positive quantity, in contrast to that for a virtual image. We also see that

> Real images form on the side of a mirror where the object is, and virtual images form on the opposite side.

As we shall prove in Section 35-8, when light rays from an object make only small angles with the central axis of a spherical mirror and the proper sign conventions are chosen, a simple equation relates the object distance o, the image distance i, and the focal length f:

$$\frac{1}{o} + \frac{1}{i} = \frac{1}{f} \quad \text{(spherical mirror).} \tag{35-13}$$

We assume such small angles in figures such as Fig. 35-19, but for clarity the rays are drawn with exaggerated angles. With that assumption, Eq. 35-13 applies to any concave, convex, or plane mirror. For a convex or plane mirror, only a virtual image can be formed, regardless of the object's location on the central axis. As shown in the example of a convex mirror in Fig. 35-17c, the image is always on the opposite side of the mirror from the object and has the same orientation as the object.

The size of an object or image, as measured *perpendicular* to the mirror's central axis, is called the object or image *height*. Let h represent the height of the object and h' the height of the image. Then the ratio h'/h is called the **lateral magnification** m produced by the mirror. However, by convention, the lateral magnification always includes a plus sign when the image orientation is the same as that of the object and a minus sign when the image orientation is opposite that of the object (that is, upside down). For this reason, we write the formula for m as

$$|m| = \frac{h'}{h} \quad \text{(lateral magnification).} \tag{35-14}$$

TABLE 35-2
Organizing Table for Mirrors

Mirror Type	Object Location	Image Location	Type	Orientation	Sign of f	of r	of i	of m
Plane	Anywhere							
Concave	Inside F							
	Outside F							
Convex	Anywhere							

We shall soon prove that the lateral magnification can also be written as

$$m = -\frac{i}{o} \quad \text{(lateral magnification).} \tag{35-15}$$

For a plane mirror for which $i = -o$, we have $m = +1$. The magnification of 1 means that the image is the same size as the object. The plus sign means that the image and the object have the same orientation. For the concave mirror of Fig. 35-19c, $m \approx -1.5$.

Equations 35-12 through 35-15 hold for all plane mirrors, concave spherical mirrors, and convex spherical mirrors. In addition to those equations, you have been asked to absorb a lot of information about these mirrors, and you should organize it for yourself by filling in Table 35-2. Under Image Location, note whether the image is on the *same* side of the mirror as the object or on the opposite side. Under Image Type, note whether the image is *real* or *virtual*. Under Image Orientation, note whether the image has the *same* orientation as the object or is *inverted*. Under Sign, give the sign of the quantity or fill in ± if the sign is ambiguous. You will need this organization to tackle homework or examinations.

Locating Images by Drawing Rays

Figures 35-20a and b show an object O in front of a concave mirror. In general, a point on an object puts out a spray of rays in all directions. For most of these rays, we need a protractor to calculate where a ray that hits our mirror (and later in the

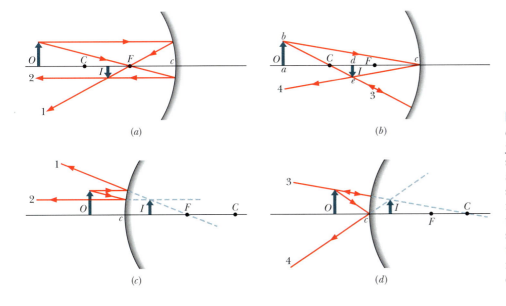

(a)

(b)

(c)

(d)

FIGURE 35-20 ■ (a, b) Four rays that can easily be drawn to find the image of an object in a concave mirror. For the object position shown, the image is real, inverted, and smaller than the object. (c, d) Four similar rays for the case of a convex mirror. For a convex mirror, the image is always virtual, oriented like the object, and smaller than the object. [In (c), ray 2 is initially directed toward focal point F. In (d), ray 3 is initially directed toward center of curvature C.]

chapter, our lens) will go. But for four special rays, we can easily draw where they are going to go, using the focal point and symmetry. We can graphically locate the image of any off-axis point of the object by drawing a *ray diagram* with any two of four special rays through the point:

1. A ray that is initially parallel to the central axis reflects through the focal point *F* (ray 1 in Fig. 35-20*a*).

2. A ray that reflects from the mirror after passing through the focal point emerges parallel to the central axis (ray 2 in Fig. 35-20*a*).

3. A ray that reflects from the mirror after passing through the center of curvature *C* returns along itself (ray 3 in Fig. 35-20*b*).

4. A ray that reflects from the mirror at its intersection *c* with the central axis is reflected symmetrically about that axis (ray 4 in Fig. 35-20*b*).

The image of the point is at the intersection of the two special rays you choose. The image of the object can then be found by locating the images of two or more of its off-axis points. You need to modify the descriptions of the rays slightly to apply them to convex mirrors, as in Figs. 35-20*c* and *d*. By referring to Figs. 35-20*c* and *d,* you can easily write a description of what happens to rays 1, 2, 3, and 4 as they are reflected from a convex mirror.

Proof of Equation 35-15

We are now in a position to derive Eq. 35-15 ($m = -i/o$), the equation for the lateral magnification of an object reflected in a mirror. Consider ray 4 in Fig. 35-20*b*. It is reflected at point *c* so that the incident and reflected rays make equal angles with the axis of the mirror at that point.

The two right triangles *abc* and *cde* in the figure are similar, so we can write

$$\frac{de}{ab} = \frac{cd}{ca}.$$

The quantity on the left (apart from the question of sign) is the lateral magnification *m* produced by the mirror. Since we indicate an inverted image as a *negative* magnification, we symbolize this as $-m$. However, $cd = i$ and $ca = o$, so we have at once

$$m = -\frac{i}{o} \quad \text{(magnification)}, \tag{35-16}$$

which is the relation we set out to prove.

READING EXERCISE 35-4: Use Figs. 35-9*c* and *d* to modify the rates developed for a concave mirror to describe what happens to rays 1, 2, 3, and 4 when they are incident on a convex mirror. ∎

READING EXERCISE 35-5: A Central American vampire bat, dozing on the central axis of a spherical mirror, is magnified by $m = -4$. Is its image (a) real or virtual, (b) inverted or of the same orientation as the bat, and (c) on the same side of the mirror as the bat or on the opposite side? ∎

TOUCHSTONE EXAMPLE 35-3: Tarantula

A tarantula of height *h* sits cautiously before a spherical mirror whose focal length has absolute value $|f| = 40$ cm. The image of the tarantula produced by the mirror has the same orientation as the tarantula and has height $h' = 0.20h$.

(a) Is the image real or virtual, and is it on the same side of the mirror as the tarantula or the opposite side?

SOLUTION ■ The **Key Idea** here is that because the image has the same orientation as the tarantula (the object), it must be virtual and on the opposite side of the mirror. (You can easily see this result if you have filled out Table 35-2).

(b) Is the mirror concave or convex, and what is its focal length f, sign included?

SOLUTION ■ We *cannot* tell the type of mirror from the type of image, because both types of mirror can produce virtual images. Similarly, we cannot tell the type of mirror from the sign of the focal length f, as obtained from Eq. 35-12 or 35-13, because we lack enough information to use either equation. However—and this is the **Key Idea** here—we can make use of the magnification information. We know that the ratio of image height h' to object height h is 0.20. Thus, from Eq. 35-14 we have

$$|m| = \frac{h'}{h} = 0.20.$$

Because the object and image have the same orientation, we know that m must be positive: $m = +0.20$. Substituting this into Eq. 35-15 and solving for, say, i gives us

$$i = -0.20o,$$

which does not appear to be of help in finding f. However, it is helpful if we substitute it into Eq. 35-13. That equation gives us

$$\frac{1}{f} = \frac{1}{i} + \frac{1}{o} = \frac{1}{-0.20o} + \frac{1}{o} = \frac{1}{o}(-5 + 1),$$

from which we find

$$f = -o/4.$$

Now we have it: Because o is positive, f must be negative, which means that the mirror is convex with

$$f = -40 \text{ cm.} \qquad \text{(Answer)}$$

35-9 Spherical Refracting Surfaces

We now turn from images formed by reflections to images formed by refraction through smooth surfaces of transparent materials, such as glass. We've seen that curved mirrors scatter light rays so that our eyes see them in new and different ways—in different places, of different sizes, and perhaps upside down. But the most powerful applications of the bending of light rays come when we consider the effect of the refraction of rays passing through transparent materials. We can then understand how the human eye works and construct optical devices with which we can look at objects and bring them into focus (eye glasses), make them bigger (microscopes and telescopes), or create images for storing (cameras). We shall consider only spherical surfaces, with radius of curvature r and center of curvature C. The light will be emitted by a point object O in a medium with index of refraction n_1; it will refract through a spherical surface into a medium of index of refraction n_2.

To determine where the image forms (that is, whether it is real or virtual), we need to once again consider the change in direction of the rays of light that strike the refracting surface. However, as opposed to image formation by mirrors where we are concerned with reflected rays, we are concerned here with the ray that enters the refracting material and is bent (refracted). The final answer to whether the image is virtual (assuming that an observer intercepts the rays) or real (no observer necessary) depends on the relative values of n_1 and n_2 and on the geometry of the situation. Specifically, to understand image formation by spherical refracting surfaces, we draw a normal to the surface. Then,

This insect has been entombed in amber for about 25 million years. Because we view the insect through a curved refracting surface, the image we see does not coincide with the insect.

The refracted ray is in the plane determined by the incident ray and the normal. The relationship between the angle of refraction and the angle of incidence is represented in the law of refraction (or Snell's law): $n_1 \sin \theta_1 = n_2 \sin \theta_2$ (Eq. 35-2).

Six possible results are shown in Fig. 35-21. In each part of the figure, the medium with the greater index of refraction is shaded, and an object O is located on a central axis passing through the center of curvature of the refracting surface. O is always in

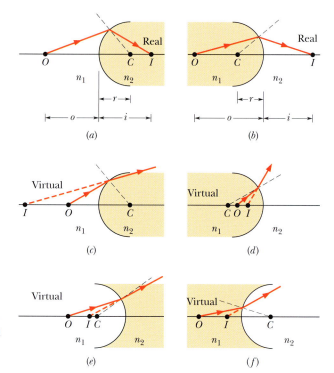

FIGURE 35-21 ■ Six possible ways in which an image can be formed by refraction through a spherical surface of radius r and center of curvature C. The surface separates a medium with index of refraction n_1 from a medium with index of refraction n_2. The point object O is always in the medium with n_1, to the left of the surface. The material with the lesser index of refraction is unshaded (think of it as being air, and the other material as being glass). Real images are formed in (a) and (b); virtual images are formed in the other four situations.

the medium with index of refraction n_1, to the left of the refracting surface. In each part, a representative ray is shown refracting through the surface. Another ray along the axis has $\theta_1 = 0$, which means $\theta_2 = 0$, so it is undeviated. The undeviated ray on the central axis and the ray refracting from an off-axis surface point along the central axis and suffice to determine the position of the image in each case.

At the point of refraction of each ray, the normal to the refracting surface is a radial line through the center of curvature C. Because of the refraction, the ray bends toward the normal if it is entering a medium of greater index of refraction, and away from the normal if it is entering a medium of lesser index of refraction. If the refracted ray is then directed toward the central axis, it and other (undrawn) rays will form a real image on that axis. If it is directed away from the central axis, it cannot form a real image; however, backward extensions of it and other refracted rays can form a virtual image, provided (as with mirrors) some of those rays are intercepted by an observer.

Real images I are formed (at image distance i) in parts a and b of Fig. 35-21, where the refraction directs the ray *toward* the central axis. Virtual images are formed in parts c and d, where the refraction directs the ray *away* from the central axis. Note that, in these four parts, real images are formed when the object is relatively far from the refracting surface, and virtual images are formed when the object is nearer the refracting surface. In the final situations (Figs. 35-21e and f), refraction always directs the ray away from the central axis and virtual images are always formed, regardless of the object distance.

Note the following major difference from reflected images:

> For a single spherical refracting surface, real images form on the side of a refracting surface that is opposite the object, and virtual images form on the same side as the object.

In Section 35-12, we shall show that (for light rays making only small angles with the central axis)

$$\frac{n_1}{o} + \frac{n_2}{i} = \frac{n_2 - n_1}{r}. \tag{35-17}$$

Just as with mirrors, the object distance o is positive, and the image distance i is positive for a real image and negative for a virtual image. However, to keep all the signs correct in Eq. 35-17, we must use the following rule for the sign of the radius of curvature r:

> When the object faces a convex refracting surface, the radius of curvature r is positive. When it faces a concave surface, r is negative.

Be careful: This is just the reverse of the sign convention we have for mirrors. Figure 35-20a shows the case of a image formation by a mirror in which the values of o, i, r, and f are all positive. Figure 35-21a shows the case of image formation by a lens in which these values are all positive. Also, don't forget that we write o, which is a distance, differently than O, which is a point, and 0, which is zero. Be careful not to confuse these quantities.

READING EXERCISE 35-6: A bee is hovering in front of the concave spherical refracting surface of a glass sculpture. (a) Which of the general situations of Fig. 35-21 is like this situation? (b) Is the image produced by the surface real or virtual, and is it on the same side as the bee or the opposite side? ∎

TOUCHSTONE EXAMPLE 35-4: Jurassic Mosquito

A Jurassic mosquito is discovered embedded in a chunk of amber, which has index of refraction 1.6. One surface of the amber is spherically convex with radius of curvature 3.0 mm (Fig. 35-22). The mosquito head happens to be on the central axis of that surface and, when viewed along the axis, appears to be buried 5.0 mm into the amber. How deep is it really?

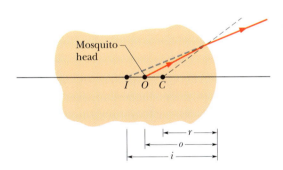

FIGURE 35-22 ∎ A piece of amber with a mosquito from the Jurassic period, with the head buried at point O. The spherical refracting surface at the right end, with center of curvature C, provides an image I to an observer intercepting rays from the object at O.

SOLUTION ∎ The **Key Idea** here is that the head only appears to be 5.0 mm into the amber because the light rays that the observer intercepts are bent by refraction at the convex amber surface. The image distance i differs from the actual object distance o according to Eq. 35-17. To use that equation to find the actual object distance, we first note:

1. Because the object (the head) and its image are on the same side of the refracting surface, the image must be virtual and so $i = -5.0$ mm.

2. Because the object is always taken to be in the medium of index of refraction n_1, we must have $n_1 = 1.6$ and $n_2 = 1.0$.

3. Because the object faces a concave refracting surface, the radius of curvature r is negative and so $r = -3.0$ mm.

Making these substitutions in Eq. 35-17,

$$\frac{n_1}{o} + \frac{n_2}{i} = \frac{n_2 - n_1}{r},$$

yields

$$\frac{1.6}{o} + \frac{1.0}{-5.0 \text{ mm}} = \frac{1.0 - 1.6}{-3.0 \text{ mm}}$$

and

$$o = 4.0 \text{ mm.} \qquad \text{(Answer)}$$

35-10 Thin Lenses

A **lens** is a transparent object with two refracting surfaces whose central axes coincide. The common central axis is the central axis of the lens. When a lens is surrounded by air, light refracts from the air into the lens, crosses through the lens, and then refracts back into the air. Each refraction can change the direction of travel of the light.

A lens that causes light rays initially parallel to the central axis to converge is (reasonably) called a **converging lens.** If, instead, it causes such rays to diverge, the lens is a **diverging lens.** When an object is placed in front of a lens of either type, refraction by the lens's surface of light rays from the object can produce an image of the object.

We shall consider only the special case of a **thin lens**—that is, a lens in which the thickest part is thin compared to the object distance o, the image distance i, and the radii of curvature r_1 and r_2 of the two surfaces of the lens. We shall also consider only light rays that make small angles with the central axis (they are exaggerated in the figures here). In Section 35-12 we shall prove that for such rays, a thin lens has a focal length f. Moreover, i and o are related to each other by

$$\frac{1}{f} = \frac{1}{o} + \frac{1}{i} \quad \text{(thin lens)}, \tag{35-18}$$

which is the same form of equation we had for mirrors. We shall also prove that when a thin lens with index of refraction n is surrounded by air, this focal length f is given by

$$\frac{1}{f} = (n-1)\left(\frac{1}{r_1} - \frac{1}{r_2}\right) \quad \text{(thin lens in air)}, \tag{35-19}$$

which is often called the *lens maker's equation.* Here r_1 is the radius of curvature of the lens surface nearer the object, and r_2 is that of the other surface. The signs of these radii are found with the rules in Section 35-9 for the radii of spherical refracting surfaces. If the lens is surrounded by some medium other than air (say, corn oil) with index of refraction n_{medium}, we replace n in Eq. 35-19 with n/n_{medium}. Keep in mind the basis of Eqs. 35-18 and 35-19:

A lens can produce an image of an object only if it bends light rays, but it can bend light rays only if its index of refraction differs from that of the surrounding medium.

Figure 35-23a shows a thin lens with convex outer surfaces, or *sides*. Once again, we figure out how the rays of light will be bent by taking a small part of the lens, near where the ray hits, and treating it as flat surface. Then, using the law of refraction (Snell's law, Eq. 35-2) which tells us $n_1 \sin \theta_1 = n_2 \sin \theta_2$, we know how the rays will be bent.

When rays that are parallel to the central axis of the lens are sent through the lens, they refract twice, as is shown enlarged in Fig. 35-23b. This double refraction causes the rays to converge and pass through a common point F_2 at a distance f from the center of the lens. Hence, this lens is a converging lens; further, a *real* focal point (or focus) exists at F_2 (because the rays really do pass through it), and the associated focal length is f. When rays parallel to the central axis are sent in the opposite direction through the lens, we find another real focal point at F_1 on the other side of the lens. For a thin lens, these two focal points are equidistant from the lens.

Because the focal points of a converging lens are real, we take the associated focal lengths f to be positive, just as we do with a real focus of a concave mirror. However, signs in optics can be tricky, so we had better check this in Eq. 35-19. The left side of that equation is positive if f is positive; how about the right side? We examine it term by term. Because the index of refraction n of glass or any other material is greater than 1, the term $(n-1)$ must be positive. Because the source of the light (which is the object) is at the left and faces the convex left side of the lens, the

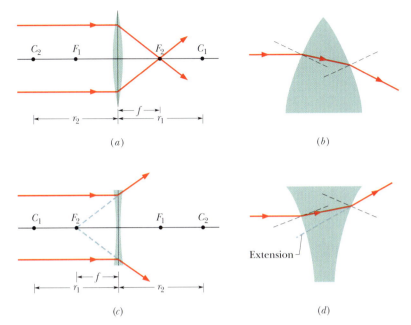

FIGURE 35-23 ■ (a) Rays initially parallel to the central axis of a converging lens are made to converge to a real focal point F_2 by the lens. The lens is thinner than drawn, with a width like that of the vertical line through it, where we shall consider all the bending of rays to occur. (b) An enlargement of the top part of the lens of (a); normals to the surfaces are shown dashed. Note that both refractions of the ray at the surfaces bend the ray downward, toward the central axis. (c) The same initially parallel rays are made to diverge by a diverging lens. Extensions of the diverging rays pass through a virtual focal point F_2. (d) An enlargement of the top part of the lens of (c). Note that both refractions of the ray at the surfaces bend the ray upward, away from the central axis.

radius of curvature r_1 of that side must be positive according to the sign rule for refracting surfaces. Similarly, because the object faces a concave right side of the lens, the radius of curvature r_2 of that side must be negative. Thus, the term $(1/r_1 - 1/r_2)$ is positive, the whole right side of Eq. 35-19 is positive, and all the signs are consistent.

Figure 35-23c shows a thin lens with concave outer surfaces. When rays that are parallel to the central axis of the lens are sent through this lens, they refract twice, as is shown enlarged in Fig. 35-23d; these rays *diverge*, never passing through any common point, and so this lens is a diverging lens. However, extensions of the rays do pass through a common point F_2 at a distance f from the center of the lens. Hence, the lens has a *virtual* focal point at F_2. (If your eye intercepts some of the diverging rays, you perceive a bright spot to be at F_2, as if it is the source of the light.) Another virtual focus exists on the opposite side of the lens at F_1, symmetrically placed if the lens is thin. Because the focal points of a diverging lens are virtual, we take the focal length f to be negative. *Note:* For a thin lens, the two focal points are at the same distance from the lens on either side even if the curvatures of the two sides are not equal and opposite.

Images from Thin Lenses

We now consider the types of image formed by converging and diverging lenses. In thinking about the image formed of an object by a lens, it is important to keep in mind that the object scatters light falling on it in all directions. However, only those rays of light that fall on the lens are refracted and have their directions changed. Thus, all of the rays that are incident on the lens contribute to the image that forms.*

Figure 35-24a shows an object O outside the focal point F_1 of a converging lens. The two rays drawn in the figure show that the lens forms a real, inverted image I of the object on the side of the lens opposite the object.

A fire is being started by focusing sunlight onto newspaper by means of a converging lens made of clear ice. The lens was made by freezing water in the shallow vessel (which has a curved bottom).

* In Section 35-8, we use Eq. 35-12 to construct diagrams (Figs. 35-19 and 35-20) for spherical mirrors in which the center curvature of the mirror is replaced by its focal point. We can construct similar diagrams for thin lenses using Eq. 35-19 to determine the locations of F based on the value of f.

FIGURE 35-24 ■ (a) A real, inverted image I is formed by a converging lens when the object O is outside the focal point F_1. (b) The image I is virtual and has the same orientation as O when O is inside the focal point. (c) A diverging lens forms a virtual image I, with the same orientation as the object O, whether O is inside or outside the focal point of the lens.

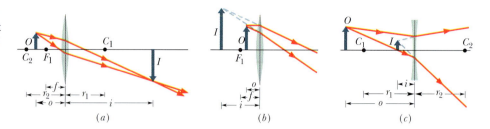

(a)   (b)   (c)

When the object is placed inside the focal point F_1, as in Fig. 35-24b, the lens forms a virtual image I on the same side of the lens as the object and with the same orientation but larger in size. In this situation, the lens acts as a magnifying glass. Hence, a converging lens can form either a real image or a virtual image, depending on whether the object is outside or inside the focal point, respectively.

Figure 35-24c shows an object O in front of a diverging lens. Regardless of the object distance (regardless of whether O is inside or outside the virtual focal point), this lens produces a virtual image that is on the same side of the lens as the object and has the same orientation.

As with mirrors, we take the image distance i to be positive when the image is real and negative when the image is virtual. However, the locations of real and virtual images from lenses are the reverse of those from mirrors:

> Real images form on the side of a lens that is opposite the object, and virtual images form on the side where the object is.

The lateral magnification m produced by converging and diverging lenses is given by Eqs. 35-14 and 35-15, the same as for mirrors.

You have been asked to absorb a lot of information in this section, and you should organize it for yourself by filling in Table 35-3 for thin lenses. Under Image Location note whether the image is on the *same* side of the lens as the object or on the *opposite* side. Under Image Type note whether the image is *real* or *virtual*. Under Image Orientation note whether the image has the same orientation as the object or is inverted.

TABLE 35-3
Organizing Table for Lenses

Lens Type	Object Location	Image Location	Image Type	Image Orientation	Sign of f	Sign of i	Sign of m
Converging	Inside F						
	Outside F						
Diverging	Anywhere						

Locating Images of Extended Objects with Principal Rays

Converging Lens: Object Outside of F_1   Figure 35-25a shows an object O outside focal point F_1 of a converging lens. We can graphically locate the image of any off-axis point on such an object (such as the tip of the arrow in Fig. 35-25a) by drawing a ray diagram with any two of three easy-to-draw principal rays through the point. These principal rays are chosen for convenience from the infinite number of rays that pass through the lens to form the image:

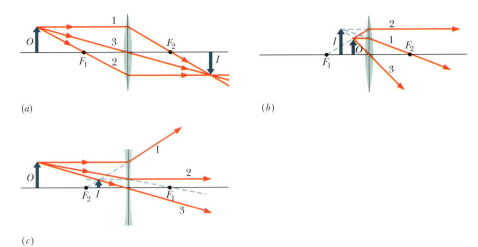

(a)

(b)

(c)

FIGURE 35-25 ■ Three special rays allow us to locate an image formed by a thin lens whether the object O is (a) outside or (b) inside the focal point of a converging lens, or (c) anywhere in front of a diverging lens.

1. A ray (from outside the focal point) that is initially parallel to the central axis of a converging lens will pass through focal point F_2 (ray 1 in Fig. 35-25a).

2. A ray that initially passes through focal point F_1 will emerge from a converging lens parallel to the central axis (ray 2 in Fig. 35-25a).

3. A ray that is initially directed toward the center of a converging lens will emerge from the lens with no change in its direction (ray 3 in Fig. 35-25a) because the ray encounters the two sides of the lens where they are almost parallel.

The image of the point is located where the rays intersect on the far side of the converging lens. The image of the object is found by locating the images of two or more of its points.

Rules for Other Situations   Figure 35-25b shows how the extensions of the three special rays can be used to locate the image of an object placed inside focal point F_1 of a converging lens. Notice that although ray 2 is determined by the focal point, no part of the ray goes through it. It is the reversal of ray 1—what ray 1 would do if it came from the other side of the lens. We suggest you develop rules like these three for Fig. 35-25a and 35-25b.

You need to modify the descriptions of rays 1 and 2 to use them to locate an image placed (anywhere) in front of a diverging lens. In Fig. 35-25c, for example, we find the intersection of ray 3 and the backward extensions of rays 1 and 2.

Two-Lens Systems

When an object O is placed in front of a system of two lenses whose central axes coincide, we can locate the final image of the system (that is, the image produced by the lens farther from the object) by working in steps. Let lens 1 be the nearer lens and lens 2 the farther lens.

Step 1.   We let o_1 represent the distance of object O from lens 1. We then find the distance i_1 of the image produced by lens 1, by use of Eq. 35-18. The image could be real or virtual.

Step 2.   Now, ignoring the presence of lens 1, we treat the image found in step 1 *as the object* for lens 2. If this new object is located beyond lens 2, the object distance o_2 for lens 2 is taken to be negative. (Note this exception to the rule that says the object distance is positive; the exception occurs because the object here is on the side opposite the source of light.) Otherwise, o_2 is taken to be positive as usual.

We then find the distance i_2 of the (final) image produced by lens 2 by use of Eq. 35-18.

A similar step-by-step solution can be used for any number of lenses or if a mirror is substituted for lens 2.

The overall lateral magnification M produced by a system of two lenses is the product of the lateral magnifications m_1 and m_2 produced by the two lenses:

$$M = m_1 m_2. \qquad (35\text{-}20)$$

TOUCHSTONE EXAMPLE 35-5: Praying Mantis

A praying mantis preys along the central axis of a thin symmetric lens, 20 cm from the lens. The lateral magnification of the mantis provided by the lens is $m = -0.25$, and the index of refraction of the lens material is 1.65.

(a) Determine the type of image produced by the lens, the type of lens, whether the object (mantis) is inside or outside the focal point, on which side of the lens the image appears, and whether the image is inverted.

SOLUTION ■ The **Key Idea** here is that we can tell a lot about the lens and the image from the given value of m. From it and Eq. 35-16 ($m = -i/o$), we see that

$$i = -(m)(o) = 0.25o.$$

Even without finishing the calculation, we can answer the questions. Because o is positive, i here must be positive. That means we have a real image, which means we have a converging lens (the only lens that can by itself produce a real image). The object must be outside the focal point (the only way a real image can be produced). Also, the image is inverted and on the side of the lens opposite the object. (That is how a converging lens makes a real image.)

(b) What are the two radii of curvature of the lens?

SOLUTION ■ The **Key Ideas** here are these:

1. Because the lens is symmetric, r_1 (for the surface nearer the object) and r_2 have the same magnitude r.

2. Because the lens is a converging lens, the object faces a convex surface on the nearer side and so $r_1 = +r$. Similarly, it faces a concave surface on the farther side and so $r_2 = -r$.

3. We can relate these radii of curvature to the focal length f via the lens maker's equation, Eq. 35-19 (our only equation involving the radii of curvature of a lens).

4. We can relate f to the object distance o and image distance i via Eq. 35-18.

We know o but we do not know i. Thus, our starting point is to finish the calculation for i in part (a); we obtain

$$i = (0.25)(20 \text{ cm}) = 5.0 \text{ cm}.$$

Now Eq. 35-18 gives us

$$\frac{1}{f} = \frac{1}{o} + \frac{1}{i} = \frac{1}{20 \text{ cm}} + \frac{1}{5.0 \text{ cm}},$$

from which we find $f = 4.0$ cm.

Equation 35-19 then gives us

$$\frac{1}{f} = (n-1)\left(\frac{1}{r_1} - \frac{1}{r_2}\right) = (n-1)\left(\frac{1}{+r} - \frac{1}{-r}\right)$$

or, with known values inserted,

$$\frac{1}{4.0 \text{ cm}} = (1.65 - 1)\frac{2}{r},$$

which yields

$$r = (0.65)(2)(4.0 \text{ cm}) = 5.2 \text{ cm}. \qquad \text{(Answer)}$$

TOUCHSTONE EXAMPLE 35-6: Jalapeño Seed

Figure 35-26a shows a jalapeño seed O_1 that is placed in front of two thin symmetrical coaxial lenses 1 and 2, with focal lengths $f_1 = +24$ cm and $f_2 = +9.0$ cm, respectively, and with lens separation $L = 10$ cm. The seed is 6.0 cm from lens 1. Where does the system of two lenses produce an image of the seed?

SOLUTION ■ We could locate the image produced by the system of lenses by tracing light rays from the seed through the two lenses. However, the **Key Idea** here is that we can, instead, calculate the location of that image by working through the system in steps, lens by lens. We begin with the lens closer to the

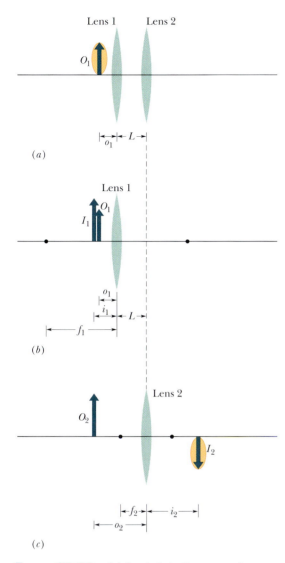

FIGURE 35-26 ■ (a) Seed O_1 is distance o_1 from a two-lens system with lens separation L. We use the arrow to orient the seed. (b) The image I_1 produced by lens 1 alone. (c) Image I_1 acts as object O_2 for lens 2 alone, which produces the final image I_2.

seed. The image we seek is the final one—that is, image I_2 produced by lens 2.

Lens 1. Ignoring lens 2, we locate the image I_1 produced by lens 1 by applying Eq. 35-18 to lens 1 alone:

$$\frac{1}{o_1} + \frac{1}{i_1} = \frac{1}{f_1}.$$

The object O_1 for lens 1 is the seed, which is 6.0 cm from the lens; thus, we substitute $o_1 = +6.0$ cm. Also substituting the given value of f_1, we then have

$$\frac{1}{+6.0 \text{ cm}} + \frac{1}{i_1} = \frac{1}{+24 \text{ cm}},$$

which yields $i_1 = -8.0$ cm.

This tells us that image I_1 is 8.0 cm to the left of lens 1 and virtual. (We could have guessed that it is virtual by noting that the seed is inside the focal point of lens 1.) Since I_1 is virtual, it is on the same side of the lens as object O_1 and has the same orientation as the seed, as shown in Fig. 35-26b.

Lens 2. In the second step of our solution, the **Key Idea** is that we can treat image I_1 as an object O_2 for the second lens and now ignore lens 1. We first note that this object O_2 is outside the focal point of lens 2. So the image I_2 produced by lens 2 must be real, inverted, and on the side of the lens opposite O_2. Let us see.

The distance o_2 between this object O_2 and lens 2 is, from Fig. 35-26c,

$$o_2 = L + |i_1| = 10 \text{ cm} + 8.0 \text{ cm} = 18 \text{ cm}.$$

Then Eq. 35-18, now written for lens 2, yields

$$\frac{1}{+18 \text{ cm}} + \frac{1}{i_2} = \frac{1}{+9.0 \text{ cm}}.$$

Hence, $\qquad\qquad i_2 = +18$ cm. (Answer)

The plus sign confirms our guess: Image I_2 produced by lens 2 is real, inverted, and on the side of lens 2 opposite O_2, as shown in Fig. 35-26c.

35-11 Optical Instruments

The human eye is a remarkably effective organ, but its range can be extended in many ways by optical instruments such as eyeglasses, simple magnifying lenses, motion picture projectors, cameras (including TV cameras), microscopes, and telescopes. Many such devices extend the scope of our vision beyond the visible range; satellite-borne infrared cameras, x-ray microscopes, and radio telescopes are examples. Furthermore, electron microscopes and magnets in particle accelerators can focus electron and proton beams.

The mirror and thin-lens formulas can be applied only as approximations to most sophisticated optical instruments. The lenses in typical laboratory microscopes are by no means "thin." In most optical instruments the lenses are compound lenses; that is, they are made of several components, the interfaces rarely being exactly spherical. A more complex treatment is needed for these systems. Now we discuss three optical instruments, assuming for simplicity, that the thin-lens formulas apply.

Simple Magnifying Lens

The normal human eye can focus a sharp image of an object on the retina (at the rear of the eye) if the object is located anywhere from infinity to a certain point called the *near point* P_n. If you move the object closer to the eye than the near point, the perceived retinal image becomes fuzzy. The location of the near point normally varies with age. We have all heard about people who claim not to need glasses but read their newspapers at arm's length; their near points are receding. To find your own near point, remove your glasses or contacts if you wear any, close one eye, and then bring this page closer to your open eye until it becomes indistinct. In what follows, we take the near point to be 25 cm from the eye, a bit more than the typical value for 20-year-olds.

Figure 35-27a shows an object O placed at the near point P_n of an eye. The size of the image of the object produced on the retina depends on the angle θ that the object occupies in the field of view from that eye. By moving the object closer to the eye, as in Fig. 35-27b, you can increase the angle and, hence, the possibility of distinguishing details of the object. However, because the object is then closer than the near point, it is no longer in focus; that is, the image is no longer clear.

You can restore the clarity by looking at O through a converging lens, placed so that O is just inside the focal point F_1 of the lens, which is at focal length f (Fig. 35-27c). What you then see is the virtual image of O produced by the lens. That image is farther away than the near point; thus, the eye can see it clearly.

Moreover, the angle θ' occupied by the virtual image is larger than the largest angle θ that the object alone can occupy and still be seen clearly. The *angular magnification* m_θ (not to be confused with lateral magnification m) of what is seen is

$$m_\theta = \theta'/\theta.$$

In words, the angular magnification of a simple magnifying lens is a comparison of the angle occupied by the image the lens produces with the angle occupied by the object when the object is moved to the near point of the viewer.

From Fig. 35-27, assuming that O is at the focal point of the lens, and approximating $\tan \theta$ as θ and $\tan \theta'$ as θ' for small angles, we have

$$\theta \approx h/25 \text{ cm} \quad \text{and} \quad \theta' \approx h/f.$$

FIGURE 35-27 ▪ (a) An object O of height h, placed at the near point of a human eye, occupies angle θ in the eye's view. (b) The object is moved closer to increase the angle, but now the observer cannot bring the object into focus. (c) A converging lens is placed between the object and the eye, with the object just inside the focal point F_1 of the lens. The image produced by the lens is then far enough away to be focused by the eye, and the image occupies a larger angle θ' than object O does in (a).

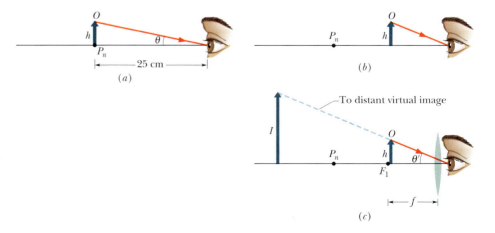

We then find that

$$m_\theta \approx \frac{25 \text{ cm}}{f} \qquad \text{(maximum magnification)}. \qquad (35\text{-}21)$$

Note that this is an equation for a person with a near point of 25 cm. It is an example of how the study of optics depends on the observer. That is, the value of the magnification depends not only on the lens but also on the person using the lens.

Compound Microscope

Figure 35-28 shows a thin-lens version of a compound microscope. The instrument consists of an *objective* (the front lens) of focal length f_{obj} and an *eyepiece* (the lens near the eye) of focal length f_{eye}. It is used for viewing small objects that are very close to the objective.

The object O to be viewed is placed just outside the first focal point F_{obj} of the objective lens, close enough to F_{obj} that we can approximate its distance o from the lens as being f_{obj}. Combining this with a consideration of the thin-lens equation

$$\frac{1}{f} = \frac{1}{o} + \frac{1}{i} \qquad \qquad (\text{Eq. } 35\text{-}18)$$

when an object under a microscope is very close to the focal point, we see that object distance o is approximately equal to focal length of the objective lens f_{obj}, so $1/f_{obj} - 1/o$ is very close to zero. As a result, the image distance i is very sensitive to *exactly how close* the object is to the focal point.

The separation between the lenses is then adjusted so that the enlarged, inverted, real image I produced by the objective is located just inside the first focal point F_1' of the eyepiece. The *tube length s* shown in Fig. 35-28 is actually large relative to f_{obj}, and we can approximate the distance i between the objective and the image I as being length s.

From Eq. 35-15, and using our approximations for o and i, we can write the lateral magnification produced by the objective as

$$m_{obj} = -\frac{i}{o} = -\frac{s}{f_{obj}}. \qquad (35\text{-}22)$$

Since the image I is located just inside the focal point F_1' of the eyepiece, the eyepiece acts as a simple magnifying lens, and an observer sees a final (virtual, inverted) image I' through it. The overall magnification of the instrument is the product of the lateral magnification m_{obj} produced by the objective (Eq. 35-22) and the angular magnification produced by the eyepiece $m_{eye} = m_\theta$ (Eq. 35-21) so that

$$M = m_{obj}m_{eye} = -\frac{s}{f_{obj}}\frac{25 \text{ cm}}{f_{eye}} \qquad \text{(microscope magnification)}. \qquad (35\text{-}23)$$

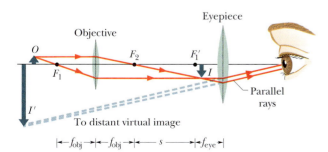

FIGURE 35-28 ■ A thin-lens representation of a compound microscope (not to scale). The objective produces a real image I of object O just inside the focal point F_1' of the eyepiece. Image I then acts as an object for the eyepiece, which produces a virtual final image I' that is seen by the observer. The objective has focal length f_{obj}; the eyepiece has focal length f_{eye}; and s is the tube length.

The microscope designer must also take into account the difference between real lenses and the ideal thin lenses we have discussed. A real lens with spherical surfaces does not form sharp images, a flaw called **spherical aberration.** Also, because refraction by the two surfaces of a real lens depends on wavelength, a real lens does not focus light of different wavelengths to the same point, a flaw called **chromatic aberration.**

Refracting Telescope

Telescopes come in a variety of forms. The form we describe here is the simple refracting telescope that consists of an objective and an eyepiece; both are represented in Fig. 35-29 with simple lenses, although in practice, as is also true for most microscopes, each lens is usually a compound-lens system to reduce distortions.

The lens arrangements for telescopes and for microscopes are similar, but telescopes are designed to view large objects, such as galaxies, stars, and planets, at large distances, whereas microscopes are designed for just the opposite purpose. This difference requires that in the telescope of Fig. 35-29 the second focal point of the objective F_2 coincide with the first focal point of the eyepiece F_1', whereas in the microscope of Fig. 35-28 these points are separated by the tube length s.

In Fig. 35-29a, parallel rays from a distant object strike the objective, making an angle θ_{obj} with the telescope axis and forming a real, inverted image at the common focal point F_2, F_1'. This image I acts as an object for the eyepiece, through which an observer sees a distant (still inverted) virtual image I'. The rays defining the image make an angle θ_{eye} with the telescope axis.

The angular magnification m_θ of the telescope is $\theta_{eye}/\theta_{obj}$. From Fig. 35-29b, for rays close to the central axis, we can write $\theta_{obj} = h'/f_{obj}$ and $\theta_{eye} \approx h'/f_{eye}$, which gives us

$$m_\theta = -\frac{f_{obj}}{f_{eye}} \qquad \text{(telescope)}, \qquad (35\text{-}24)$$

where the minus sign indicates that I' is inverted. In words, the angular magnification of a telescope is a comparison of the angle occupied by the image the telescope produces with the angle occupied by the distant object as seen without the telescope.

Magnification is only one of the design factors for an astronomical telescope and is indeed easily achieved. A good telescope needs light-gathering power, which determines how bright the image is. This is important for viewing faint objects such as distant galaxies and is accomplished by making the objective diameter as large as possible. A telescope also needs resolving power, which is the ability to distinguish

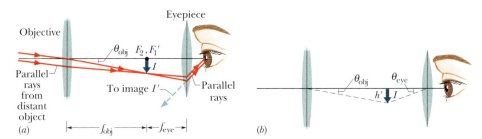

FIGURE 35-29 ■ (a) A thin-lens representation of a refracting telescope. The objective produces a real image I of a distant source of light (the object), with approximately parallel light rays at the objective. (One end of the object is assumed to lie on the central axis.) Image I, formed at the common focal points F_2 and F_1', acts as an object for the eyepiece, which produces a virtual final image I' at a great distance from the observer. The objective has focal length f_{obj}; the eyepiece has focal length f_{eye}. (b) Image I has height h' and takes up angle θ_{obj} measured from the objective and angle θ_{eye} measured from the eyepiece.

between two distant objects (stars, say) whose angular separation is small. Field of view is another important design parameter. A telescope designed to look at galaxies (which occupy a tiny field of view) is much different from one designed to track meteors (which move over a wide field of view).

The telescope designer must also take into account the differences between real lenses and the ideal thin lenses we have discussed. Designers use compound lens systems to minimize spherical and chromatic aberrations.

This brief discussion by no means exhausts the design parameters of astronomical telescopes—many others are involved. We could make a similar listing for any other high-performance optical instrument.

35-12 Three Proofs

The Spherical Mirror Formula (Eq. 35-13)

Let us prove that Eq. 35-13, $1/i + 1/o = 1/f$ is true for a spherical mirror. Figure 35-30 shows a point object O placed on the central axis of a concave spherical mirror, outside its center of curvature C. Here, we use the reflection principle that comes from treating the mirror as approximately flat near where the ray hits. A ray from O that makes an angle α with the axis intersects the axis at I after reflection from the mirror at a. A ray that leaves O along the axis is reflected back along itself at c and also passes through I. Thus, I is the image of O; it is a real image because light actually passes through it. Let us find the image distance i.

A trigonometry theorem that is useful here tells us that an exterior angle of a triangle is equal to the sum of the two opposite interior angles. Applying this to triangles OaC and OaI in Fig. 35-30 yields

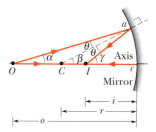

FIGURE 35-30 ■ A concave spherical mirror forms a real point image I by reflecting light rays from a point object O.

$$\beta = \alpha + \theta \quad \text{and} \quad \gamma = \alpha + 2\theta.$$

If we eliminate θ between these two equations, we find

$$\alpha + \gamma = 2\beta. \tag{35-25}$$

We can write angles α, β, and γ, in radian measure, as

$$\alpha \approx \frac{\widehat{ac}}{cO} = \frac{\widehat{ac}}{o}, \qquad \beta = \frac{\widehat{ac}}{cC} = \frac{\widehat{ac}}{r},$$

and

$$\gamma \approx \frac{\widehat{ac}}{cI} = \frac{\widehat{ac}}{i}. \tag{35-26}$$

Only the equation for β is exact, because the center of curvature of arc ac is at C. Here, $\widehat{ac}$ is the arc extending from the point that the ray reflects from the mirror to the central axis. However, the equations for α and γ are approximately correct if these angles are small enough (that is, for rays close to the central axis). Substituting Eqs. 35-26 into Eq. 35-25, using Eq. 35-12 to replace r with $2f$, and canceling $\widehat{ac}$ lead exactly to Eq. 35-13, the relation that we set out to prove.

The Refracting Surface Formula (Eq. 35-17)

Next we prove that when light from an object passes through a medium having an index of refraction n_1 and encounters a smooth spherical refracting surface of a

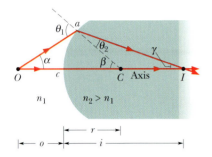

FIGURE 35-31 ■ A real point image I of a point object O is formed by refraction at a spherical convex surface between two media.

medium with an index of refraction n_2, the image and object distances are related by Eq. 35-17,

$$\frac{n_1}{o} + \frac{n_2}{i} = \frac{n_1 - n_2}{r}.$$

Let's start by considering how an incident ray from point object O in Fig. 35-31 is refracted. According to Eq. 35-2, $n_1 \sin \theta_1 = n_2 \sin \theta_2$.

If α is small, θ_1 and θ_2 will also be small and we can replace the sines of these angles with the angles themselves. Thus, the equation above becomes

$$n_1 \theta_1 \approx n_2 \theta_2. \tag{35-27}$$

We again use the fact that an exterior angle of a triangle is equal to the sum of the two opposite interior angles. Applying this to triangles COa and ICa yields

$$\theta_1 = \alpha + \beta \quad \text{and} \quad \beta = \theta_2 + \gamma. \tag{35-28}$$

If we use Eqs. 35-28 to eliminate θ_1 and θ_2 from Eq. 35-27, we find

$$n_1 \alpha + n_2 \gamma = (n_2 - n_1)\beta. \tag{35-29}$$

In radian measure the angles α, β, and γ, are

$$\alpha \approx \frac{\widehat{ac}}{o}; \qquad \beta = \frac{\widehat{ac}}{r}; \qquad \gamma \approx \frac{\widehat{ac}}{i}. \tag{35-30}$$

Only the second of these equations is exact. The other two are approximate because I and O are not the centers of circles of which ac is a part. However, for α small enough (for rays close to the axis), the inaccuracies in Eqs. 35-30 are small. Substituting Eqs. 35-30 into Eq. 35-29 leads directly to Eq. 35-17, the relation we set out to prove.

The Thin-Lens Formulas (Eqs. 35-18 and 35-19)

Finally, we set out to show that for a thin lens, the object and image distances are related to the focal length of the lens by Eq. 35-18, $1/o + 1/i = 1/f$. Our plan is to consider each lens surface as a separate refracting surface, and to use the image formed by the first surface as the object for the second.

We start with the thick glass "lens" of length L in Fig. 35-32a whose left and right refracting surfaces are ground to radii r' and r''. A point object O' is placed near the left surface as shown. A ray leaving O' along the central axis is not deflected on entering or leaving the lens.

A second ray leaving O' at an angle α with the central axis intersects the left surface at point a', is refracted, and intersects the second (right) surface at point a''. The ray is again refracted and crosses the axis at I'', which, being the intersection of two rays from O', is the image of point O', formed after refraction at two surfaces.

Figure 35-32b shows that the first (left) surface also forms a virtual image of O' at I'. To locate I', we use Eq. 35-17,

$$\frac{n_1}{o'} + \frac{n_2}{i'} = \frac{n_2 - n_1}{r'}.$$

Putting $n_1 = 1$ for air and $n_2 = n$ for lens glass and bearing in mind that the image distance is negative (that is, $i = -i'$ in Fig. 35-32b), we obtain

$$\frac{1}{o'} - \frac{n}{|i'|} = \frac{n-1}{r'}. \tag{35-31}$$

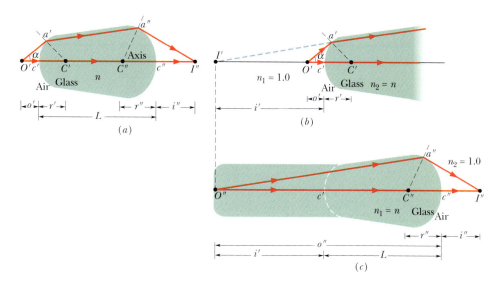

FIGURE 35-32 ■ (*a*) Two rays from point object *O'* form a real image *I"* after refracting through two spherical surfaces of a "lens." The object faces a convex surface at the left side of the lens and a concave surface at the right side. The ray traveling through points *a'* and *a"* is actually close to the central axis through the lens. (*b*) The left side and (*c*) the right side of the "lens" in (*a*), shown separately.

Figure 35-32*c* shows the second surface again. Unless an observer at point *a"* were aware of the existence of the first surface, the observer would think that the light striking that point originated at point *I'* in Fig. 35-32*b* and that the region to the left of the surface was filled with glass as indicated. Thus, the (virtual) image *I'* formed by the first surface serves as a real object *O"* for the second surface. The distance of this object from the second surface is

$$o'' = |i'| + L. \tag{35-32}$$

To apply Eq. 35-17 to the second surface, we must insert $n_1 = n$ and $n_2 = 1$ because the object now is effectively imbedded in glass. If we substitute with Eq. 35-31, then Eq. 35-17 becomes

$$\frac{n}{|i'| + L} + \frac{1}{i''} = \frac{1 - n}{r''}. \tag{35-33}$$

Let us now assume that the thickness *L* of the "lens" in Fig. 35-32*a* is so small that we can neglect it in comparison with our other linear quantities (such as o', i', o'', i'', r', and r''). In all that follows we make this *thin-lens approximation*. Putting $L = 0$ in Eq. 35-33 and rearranging the right side lead to

$$\frac{n}{i'} + \frac{1}{i''} = -\frac{n - 1}{r''}. \tag{35-34}$$

Adding Eqs. 35-31 and 35-34 leads to

$$\frac{1}{o'} + \frac{1}{i''} = (n - 1)\left(\frac{1}{r'} - \frac{1}{r''}\right).$$

Finally, calling the original object distance simply *o* and the final image distance simply *i* leads to

$$\frac{1}{o} + \frac{1}{i} = (n - 1)\left(\frac{1}{r'} - \frac{1}{r''}\right), \tag{35-35}$$

which, with $r' \equiv r_1$ and $r'' = r_2$, reduces to Eqs. 35-18 and 35-19, the relations we set out to prove.

Problems

SEC. 35-2 ■ REFLECTION AND REFRACTION

1. Light in a Vacuum Light in vacuum is incident on the surface of a glass slab. In the vacuum the beam makes an angle of 32.0° with the normal to the surface, while in the glass it makes an angle of 21.0° with the normal. What is the index of refraction of the glass?

2. Two Perpendicular Surfaces Figure 35-33 shows light reflecting from two perpendicular reflecting surfaces A and B. Find the angle between the incoming ray i and the outgoing ray r'.

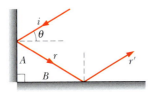

FIGURE 35-33 ■ Problem 2.

3. Rectangular Metal Tank When the rectangular metal tank in Fig. 35-34 is filled to the top with an unknown liquid, an observer with eyes level with the top of the tank can just see the corner E; a ray that refracts toward the observer at the top surface of the liquid is shown. Find the index of refraction of the liquid.

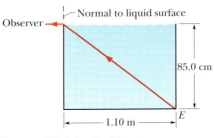

FIGURE 35-34 ■ Problem 3.

4. Claudius Ptolemy In about A.D. 150, Claudius Ptolemy gave the following measured values for the angle of incidence θ_1 and the angle of refraction θ_2 for a light beam passing from air to water:

θ_1	θ_2	θ_1	θ_2
10°	8°00′	50°	35°00′
20°	15°30′	60°	45°30′
30°	22°30′	70°	45°30′
40°	29°00′	80°	50°00′

(a) Are these data consistent with the law of refraction? (b) If so, what index of refraction results? These data are interesting as perhaps the oldest recorded physical measurements.

5. Vertical Pole In Fig. 35-35, a 2.00-m-long vertical pole extends from the bottom of a swimming pool to a point 50.0 cm above the water. Sunlight is incident at 55.0° above the horizon. What is the length of the shadow of the pole on the level bottom of the pool?

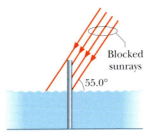

FIGURE 35-35 ■ Problem 5.

6. Four Transparent Materials In Fig. 35-36, light is incident at angle $\theta_1 = 40.1°$ on a boundary between two transparent materials. Some of the light then travels down through the next three layers of transparent materials, while some of it reflects upward and then escapes into the air. What are the values of (a) θ_5 and (b) θ_4?

7. Sideways Displacement Prove that a ray of light incident on the surface of a sheet of plate glass of thickness t emerges from the opposite face parallel to its initial direction but displaced sideways, as in Fig. 35-37. Show that, for small angles of incidence θ, this displacement is given by

$$x = t\theta\frac{n-1}{n},$$

where n is the index of refraction of the glass and θ is measured in radians.

8. White Light A ray of white light makes an angle of incidence of 35° on one face of a prism of fused quartz; the prism's cross section is an equilateral triangle. Sketch the light as it passes through the prism, showing the paths traveled by rays representing (a) blue light, (b) yellow-green light, and (c) red light.

9. Triangular Prism In Fig. 35-38, a ray is incident on one face of a triangular glass prism in air. The angle of incidence θ is chosen so that the emerging ray also makes the same angle θ with the normal to the other face. Show that the index of refraction n of the glass prism is given by

$$n = \frac{\sin\frac{1}{2}(\psi + \phi)}{\sin\frac{1}{2}\phi},$$

where φ is the vertex angle of the prism and ψ is the *deviation angle,* the total angle through which the beam is turned in passing through the prism. (Under these conditions the deviation angle ψ has the smallest possible value, which is called the *angle of minimum deviation.*)

10. Perpendicular Mirrors In Fig. 35-39 two perpendicular mirrors form the sides of a vessel filled with

FIGURE 35-36 ■ Problem 6.

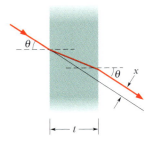

FIGURE 35-37 ■ Problem 7.

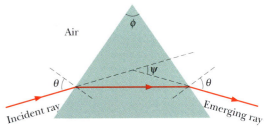

FIGURE 35-38 ■ Problems 9 and 18.

FIGURE 35-39 ■ Problem 10.

water. (a) A light ray is incident from above, normal to the water surface. Show that the emerging ray is parallel to the incident ray. Assume that there are reflections at both mirror surfaces. (b) Repeat the analysis for the case of oblique incidence with the incident ray in the plane of the figure.

SEC. 35-3 ■ TOTAL INTERNAL REFLECTION

11. Glass Slab In Fig. 35-40 a light ray enters a glass slab at point A and then undergoes total internal reflection at point B. What minimum value for the index of refraction of the glass can be inferred from this information?

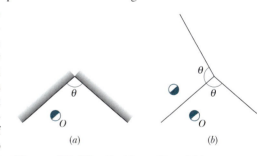

FIGURE 35-40 ■ Problem 11.

12. Benzene The index of refraction of benzene is 1.8. What is the critical angle for a light ray traveling in benzene toward a plane layer of air above the benzene?

13. Perpendicular to Face In Fig. 35-41, a ray of light is perpendicular to the face ab of a glass prism ($n = 1.52$). Find the largest value of the angle ϕ so that the ray is totally reflected at face ac if the prism is immersed (a) in air and (b) in water.

FIGURE 35-41 ■ Problem 13.

14. Point Source A point source of light is 80.0 cm below the surface of a body of water. Find the diameter of the circle at the surface through which light emerges from the water.

15. Solid Glass Cube A solid glass cube, of edge length 10 mm and index of refraction 1.5, has a small spot at its center. (a) What parts of each cube face must be covered to prevent the spot from being seen, no matter what the direction of viewing? (Neglect light that reflects inside the cube and then refracts out into the air.) (b) What fraction of the cube surface must be so covered?

16. Fused Quartz A ray of white light travels through fused quartz that is surrounded by air. If all the color components of the light undergo total internal reflection at the surface, then the reflected light forms a reflected ray of white light. However, if the color component at one end of the visible range (either blue or red) partially refracts through the surface into the air, there is less of that component in the reflected light. Then the reflected light is not white but has the tint of the opposite end of the visible range. (If blue were partially lost to refraction, then the reflected beam would be reddish, and vice versa.) Is it possible for the reflected light to be (a) bluish of (b) reddish? (c) If so, what must be the angle of incidence of the original white light on the quartz surface? (See Fig. 35-31.)

17. 90° Prism In Fig. 35-42, light enters a 90° trianglular prism at point P with incident angle θ and then some of it refracts at point Q with an angle of refraction of 90°. (a) What is the index of refraction of the prism in terms of θ? (b) What,

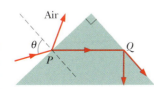

FIGURE 35-42 ■ Problem 17.

numerically, is the maximum value that the index of refraction can have? Explain what happens to the light at Q if the incident angle at Q is (c) increased slightly and (d) decreased slightly.

18. Apex Angle Given Suppose the prism of Fig. 35-38 has apex angle $\phi = 60.0°$ and index of refraction $n = 1.60$. (a) What is smallest angle of incidence θ for which a ray can enter the left face of the prism and exit the right face? (b) What angle of incidence θ is required for the ray to exit the prism with an identical angle θ for its refraction, as it does in Fig. 35-38? (See Problem 9.)

SEC. 35-4 ■ POLARIZATION BY REFLECTION

19. Light in Water Light traveling in water of refractive index 1.33 is incident on a plate of glass with index of refraction 1.53. At what angle of incidence is the reflected light fully polarized?

20. Completely Polarized (a) At what angle of incidence will the light reflected from water be completely polarized? (b) Does this angle depend on the wavelength of the light?

SEC. 35-6 ■ PLANE MIRRORS

21. Moth A moth at about eye level is 10 cm in front of a plane mirror; you are behind the moth, 30 cm from the mirror. What is the distance between your eyes and the apparent position of the moth's image in the mirror?

22. Hummingbird You look through a camera toward an image of a hummingbird in a plane mirror. The camera is 4.30 m in front of the mirror. The bird is at camera level, 5.00 m to your right and 3.30 m from the mirror. What is the distance between the camera and the apparent position of the bird's image in the mirror?

23. Two Vertical Mirrors Figure 35-43a is an overhead view of two vertical plane mirrors with an object O placed between them. If you look into the mirrors, you see multiple images of O. You can find them by drawing the reflection in each mirror of the angular region between the mirrors, as is done for the left-hand mirror in Fig. 35-43b. Then draw the reflection of the reflection. Continue this on the left and on the right until the reflections meet or overlap at the rear of the mirrors. Then you can count the number of images of O. (a) If $\theta = 90°$, how many images of O would you see? (b) Draw their locations and orientations (as in Fig. 35-43b).

FIGURE 35-43 ■ Problems 23 and 24.

24. Repeat Repeat Problem 23 for the mirror angle θ equal to (a) 45°, (b) 60°, and (c) 120°. (d) Explain why there are several possible answers for (c).

25. Prove That Prove that if a plane mirror is rotated through an angle α, the reflected beam is rotated through an angle 2α. Show that this result is reasonable for $\alpha = 45°$.

26. Corridor Figure 35-44 shows an overhead view of a corridor with a plane mirror M mounted at one end. A burglar B sneaks along the corridor directly toward the center of the mirror. If $d = 3.0$ m, how far from the mirror will she be when the security guard S can first see her in the mirror?

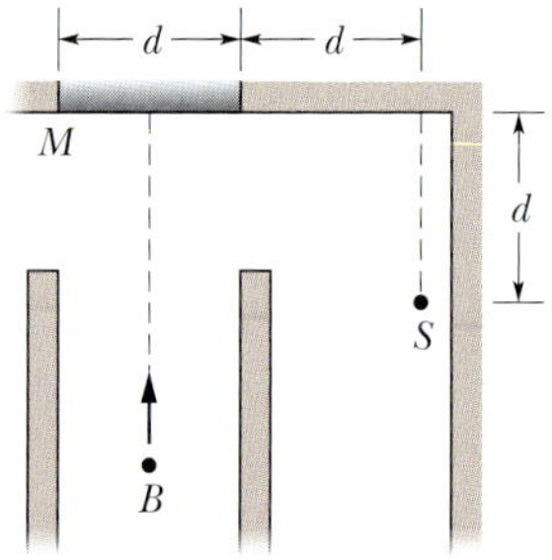

27. S and d You put a point source of light S a distance d in front of a screen A. How is the light intensity at the center of the screen changed if you put a completely reflecting mirror M a distance d behind the source, as in Fig. 35-45? (*Hint:* Use Eq. 34-25.)

28. Small Lightbulb Figure 35-46 shows a small lightbulb suspended above the surface of the water in a swimming pool. The bottom of the pool is a large mirror. How far below the mirror's surface is the image of the bulb? (*Hint:* Construct a diagram of two rays like that of Fig. 35-14, but take into account the bending of light rays by refraction. Assume that the rays are close to a vertical axis through the bulb, and use the small-angle approximation that $\sin \theta \approx \tan \theta$.)

SEC. 35-8 ■ IMAGES FROM SPHERICAL MIRRORS

29. Concave Shaving Mirror A concave shaving mirror has a radius of curvature of 35.0 cm. It is positioned so that the (upright) image of a man's face is 2.50 times the size of the face. How far is the mirror from the face?

30. Fill in Table Fill in Table 35-4, each row of which refers to a different combination of an object and either a plane mirror, a spherical convex mirror, or a spherical concave mirror. Distances are in centimeters. If a number lacks a sign, find the sign. Sketch each combination and draw in enough rays to locate the object and its image.

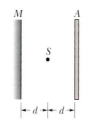

FIGURE 35-44 ■
Problem 26.

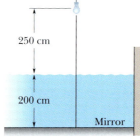

FIGURE 35-45 ■
Problem 27.

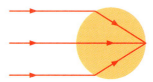

FIGURE 35-46 ■
Problem 28.

31. Short Straight Object A short straight object of length L lies along the central axis of a spherical mirror of focal length f, a distance o from the mirror. (a) Show that its image in the mirror has a length L' where

$$L' = L\left(\frac{f}{o - f}\right)^2.$$

(*Hint:* Locate the two ends of the object.) (b) Show that the *longitudinal magnification* $m' (= L'/L)$ is equal to m^2, where m is the lateral magnification.

32. Luminous Point (a) A luminous point is moving at speed v_O toward a spherical mirror with radius of curvature r, along the central axis of the mirror. Show that the image of this point is moving at speed

$$v_I = -\left(\frac{r}{2o - r}\right)^2 v_O,$$

where o is the distance of the luminous point from the mirror at any given time. (*Hint:* Start with Eq. 35-13.) Now assume that the mirror is concave, with $r = 15$ cm, and let $v_O = 5.0$ cm/s. Find the speed of the image when (b) $o = 30$ cm (far outside the focal point), (c) $o = 8.0$ cm (just outside the focal point), and (d) $o = 10$ mm (very near the mirror).

SEC. 35-9 ■ SPHERICAL REFRACTING SURFACES

33. Parallel Light Rays A beam of parallel light rays from a laser is incident on a solid transparent sphere of index of refraction n (Fig. 35-47). (a) If a point image is produced at the back of the sphere, what is the index of refraction of the sphere? (b) What index of refraction, if any, will produce a point image at the center of the sphere?

FIGURE 35-47 ■
Problem 33.

34. Fill in Table Two Fill in Table 35-5, each row of which refers to a different combination of a point object and a spherical refracting surface separating two media with different indexes of refraction. Distances are in centimeters. If a number lacks a sign, find the sign. Sketch each combination and draw in enough rays to locate the object and image.

TABLE 35-4 Problem 30: Mirrors							
Type	f	r	i	o	m	Real Image?	Inverted Image?
(a) Concave	20			+10			
(b)			+10	+1.0	No		
(c)	+20			+30			
(d)				+60	−0.50		
(e)		−40	−10				
(f)	20				+0.10		
(g) Convex		40	4.0				
(h)			+24	0.50			Yes

TABLE 35-5 Problem 34: Spherical Refracting Surfaces						
	n_1	n_2	o	i	r	Inverted Image?
(a)	1.0	1.5	+10		+30	
(b)	1.0	1.5	+10	−13		
(c)	1.0	1.5		+600	+30	
(d)	1.0		+20	−20	−20	
(e)	1.5	1.0	+10	−6.0		
(f)	1.5	1.0		−7.5	−30	
(g)	1.5	1.0	+70		+30	
(h)	1.5			+100	+600	−30

35. Coin in a Pool You look downward at a coin that lies at the bottom of a pool of liquid with depth d and index of refraction n (Fig. 35-48). Because you view with two eyes, which intercept different rays of light from the coin, you perceive the coin to be where extensions of the intercepted rays cross, at depth d_a instead of d. Assuming that the intercepted rays in Fig. 35-48 are close to a vertical axis through the coin, show that $d_a = d/n$. (*Hint:* Use the small-angle approximation that $\sin \theta \approx \tan \theta \approx \theta$.)

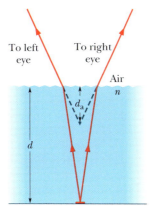

FIGURE 35-48 ■ Problem 35.

36. Carbon Tetrachloride A 20-mm-thick layer of water ($n = 1.33$) floats on a 40-mm-thick layer of carbon tetrachloride ($n = 1.46$) in a tank. A coin lies at the bottom of the tank. At what depth below the top water surface do you perceive the coin? (*Hint:* Use the result and assumptions of Problem 35 and work with a ray diagram of the situation.)

SEC. 35-10 ■ THIN LENSES

37. Thin Diverging Lens An object is 20 cm to the left of a thin diverging lens having a 30 cm focal length. What is the image distance i? Find the image position with a ray diagram.

38. Image of Sun You produce an image of the Sun on a screen, using a thin lens whose focal length is 20.0 cm. What is the diameter of the image? (See Appendix C for needed data on the Sun.)

39. Double-Convex A double-convex lens is to be made of glass with an index of refraction of 1.5. One surface is to have twice the radius of curvature of the other and the focal length is to be 60 mm. What are the radii?

40. One Side Is Flat A lens is made of glass having an index of refraction of 1.5. One side of the lens is flat, and the other is convex with a radius of curvature of 20 cm. (a) Find the focal length of the lens. (b) If an object is placed 40 cm in front of the lens, where will the image be located?

41. Newtonian Form The formula

$$\frac{1}{o} + \frac{1}{i} = \frac{1}{f}$$

is called the *Gaussian* form of the thin-lens formula. Another form of this formula, the *Newtonian* form, is obtained by considering the distance x from the object to the first focal point and the distance x' from the second focal point to the image. Show that

$$xx' = f^2$$

is the Newtonian form of the thin-lens formula.

42. Movie Camera A movie camera with a (single) lens of focal length 75 mm takes a picture of a 180-cm-high person standing 27 m away. What is the height of the image of the person on the film?

43. Illuminated Slide An illuminated slide is held 44 cm from a screen. How far from the slide must a lens of focal length 11 cm be placed to form an image of the slide's picture on the screen?

44. Fill in Table Three To the extent possible, fill in Table 35-6, each row of which refers to a different combination of an object and a thin lens. Distances are in centimeters. For the type of lens, use C for converging and D for diverging. If a number (except for the index of refraction) lacks a sign, find the sign. Sketch each combination and draw in enough rays to locate the object and image.

TABLE 35-6
Problem 44: Thin Lenses

	Type	f	r_1	r_2	i	o	n	m	Real Image?	Inverted Image?
(a)	C	10				+20				
(b)		+10				+5.0				
(c)		10				+5.0		> 1.0		
(d)		10				+5.0		< 1.0		
(e)			+30	−30		+10	1.5			
(f)			−30	+30		+10	1.5			
(g)			−30	−60		+10	1.5			
(h)						+10		0.50	No	
(i)						+10		−0.50		

45. Show That Show that the distance between an object and its real image formed by a thin converging lens is always greater than or equal to four times the focal length of the lens.

46. Diverging and Converging A diverging lens with a focal length of −15 cm and a converging lens with a focal length of 12 cm have a common central axis. Their separation is 12 cm. An object of height 1.0 cm is 10 cm in front of the diverging lens, on the common central axis. (a) Where does the lens combination produce the final image of the object (the one produced by the second, converging lens)? (b) What is the height of that image? (c) Is the image real or virtual? (d) Does the image have the same orientation as the object or is it inverted?

47. Final Image A converging lens with a focal length of +20 cm is located 10 cm to the left of a diverging lens having a focal length of −15 cm. If an object is located 40 cm to the left of the converging lens, locate and describe completely the final image formed by the diverging lens.

48. Location and Size An object is 20 cm to the left of a lens with a focal length of +10 cm. A second lens of focal length +12.5 cm is 30 cm to the right of the first lens. (a) Find the location and relative size of the final image. (b) Verify your conclusions by drawing the lens system to scale and constructing a ray diagram. (c) Is the final image real of virtual? (d) Is it inverted?

49. Two Thin Lenses Two thin lenses of focal lengths f_1 and f_2 are in contact. Show that they are equivalent to a single thin lens with

$$f = \frac{f_1 f_2}{f_1 + f_2}$$

as its focal length.

50. Real Inverted In Fig 35-49, a real inverted image I of an object O is formed by a certain lens (not shown); the object–image separation is $d = 40.0$ cm, measured along the central axis of the lens. The image is just half the size of the object. (a) What kind of lens must be used to produce this image? (b) How far from the object must the lens be placed? (c) What is the focal length of the lens?

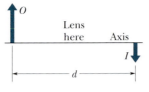

FIGURE 35-49 ■ Problem 50.

51. Object–Screen Distance A luminous object and a screen are a fixed distance D apart. (a) Show that a converging lens of focal length f, placed between object and screen, will form a real image on the screen for two lens positions that are separated by a distance

$$d = \sqrt{D(D - 4f)}.$$

(b) Show that

$$\left(\frac{D - d}{D + d}\right)^2$$

gives the ratio of the two image sizes for these two positions of the lens.

SEC. 35-11 ■ OPTICAL INSTRUMENTS

52. Astronomical Telescope If an angular magnification of an astronomical telescope is 36 and the diameter of the objective is 75 mm, what is the minimum diameter of the eyepiece required to collect all the light entering the objective from a distant point source on the telescope axis?

53. Microscope In a microscope of the type shown in Fig. 35-28, the focal length of the objective is 4.00 cm, and that of the eyepiece is 8.00 cm. The distance between the lenses is 25.0 cm. (a) What is the tube length s? (b) If image I in Fig. 35-28 is to be just inside focal point F_1', how far from the objective should the object be? What then are (c) the lateral magnification m of the objective, (d) the angular magnification m_θ of the eyepiece, and (e) the overall magnification M of the microscope?

54. Magnifying Lens A simple magnifying lens of focal length f is placed near the eye of someone whose near point P_n is 25 cm from the eye. An object is positioned so that its image in the magnifying lens appears at P_n. (a) What is the lens's angular magnification? (b) What is the angular magnification if the object is moved so that its image appears at infinity? (c) Evaluate the angular magnifications of (a) and (b) for $f = 10$ cm. (Viewing an image at P_n requires effort by muscles in the eye, whereas for many people viewing an image at infinity requires no effort.)

55. Human Eye Figure 35-50a shows the basic structure of a human eye. Light refracts into the eye through the cornea and is then further redirected by a lens whose shape (and thus ability to focus the light) is controlled by muscles. We can treat the cornea and eye lens as a single effective thin lens (Fig. 35-50b). A "normal" eye can focus parallel light rays from a distance object O to a point on the retina at the back of the eye, where processing of the visual information begins. As an object is brought close to the eye, however, the muscles must change the shape of the lens so that rays form an inverted real image on the retina (Fig. 35-50c). (a) Suppose that for the parallel rays of Figs. 35-50a and b, the focal length f of the effective thin lens of the eye is 2.50 cm. For an object at distance $o = 40.0$ cm, what focal length f' of the effective lens is required for it to be seen clearly? (b) Must the eye muscles increase or decrease the radii of curvature of the eye lens to give focal length f'?

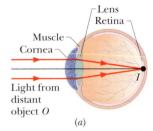

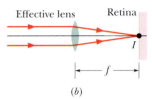

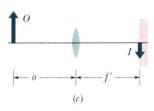

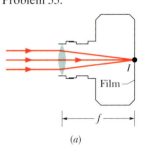

FIGURE 35-50 ■ Problem 55.

56. Compound Microscope An object is 10.0 mm from the objective of a certain compound microscope. The lenses are 300 mm apart and the intermediate image is 50.0 mm from the eyepiece. What overall magnification is produced by the instrument?

57. Camera Figure 35-51a shows the basic structure of a camera. A lens can be moved forward or back to produce an image on film at the back of the camera. For a certain camera, with the distance i between the lens and the film set at $f = 5.0$ cm, parallel light rays from a very distant object O converge to a point image on the film, as shown. The object is now brought closer, to a distance of $o = 100$ cm, and the lens–film distance is adjusted so that an inverted real image forms on the film (Fig. 35-51b). (a) What is the lens–film distance i now? (b) By how much was i changed?

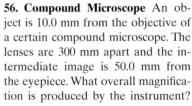

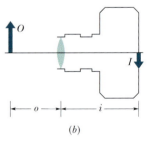

FIGURE 35-51 ■ Problem 57.

Additional Problems

58. Bizarre Behavior with Light You may have observed people with cameras behaving strangely:

(a) At a conference in North Carolina, one of the physics graduate students (who should have known better!) tried to take a picture of the overheads projected on a white screen. Since the room was darkened, he used a flash. Explain why this is a bad idea and what his pictures are likely to show.

(b) Someone mentioned to the student that he probably should not be using his flash, so he turned it off. He then proceeded to try and take pictures of the participants in the darkened room! Explain why this is a bad idea and what his pictures are likely to show.

(c) A woman on an airplane at night with a camera was impressed with the view of the city lights in the dark as the plane flew over Washington, D.C. She stood back in the aisle with her camera and tried to take a picture through the window using her flash. Explain why this is a bad idea and what her pictures are likely to show.

59. Closer Than They May Appear When a *T. rex* pursues a jeep in the movie *Jurassic Park,* we see a reflected image of the (very large) *T. rex* via a side-view mirror, on which is printed the (then darkly humorous) warning: "Objects in mirror are closer than they appear." Is the mirror flat, convex, or concave? Why do you think so?

60. Where Can You See the Bulb? In Fig. 35-52, *M* is a plane mirror; *B* is a very small bright lightbulb that can be treated as a point source of light; and *H* is an opaque housing that does not transmit light. An

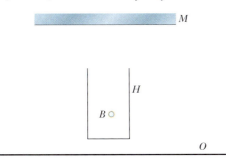

FIGURE 35-52 ■ Problem 60.

observer can stand anywhere along a line *O* to try to see the image of the lightbulb in the mirror. By using relevant rays of light, determine those locations along the line *O* from which the image of *B* is visible and those locations from which it is not visible. Mark the regions along line *O* accordingly, and explain the reasoning you used in drawing the rays. (Arons, Arnold, *A Guide to Introductory Physics Teaching,* John Wiley and Sons, New York, 1990.)

61. Who Sees What? Figure 35-53 shows a small object (represented by an arrow) in front of a curved mirror. At the tip of the arrow is a red dot. The mirror is a piece of a sphere. *The center of the sphere is marked in the picture with an x.* Eyes corresponding to three different observers are shown. For each question, explain how you got your result. Be sure to include a ray diagram as part of your explanation.

(a) How many red dots will the observer at position *A* see? Where will the dots appear to be? Specify quantitatively how far from the

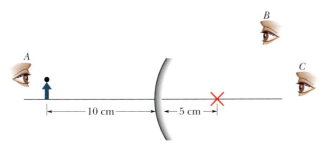

FIGURE 35-53 ■ Problem 61.

mirror the dots will appear to be and how far off the axis they will be.

(b) How many red dots will the observer at position *B* see? Explain how you know.

(c) How many red dots will the observer at position *C* see? Explain how you know.

62. The Camera and the Slide Projector Address each part of this question in two ways: (1) by drawing and interpreting appropriate geometrical diagrams and (2) by appealing to the lens equation and the expression for lateral magnification and demonstrating your result mathematically. If your two approaches do not agree, explain which one is correct and why the other is wrong.

(a) Suppose you are using a camera and wish to have a larger image of a distant object than you are obtaining with the lens currently in use. Would you change to a lens with a longer or a shorter focal length? Explain your reasoning. (*Hint:* Note that the object distance is essentially fixed.)

(b) Suppose you are using a slide projector and wish to obtain a larger image on the screen. You cannot achieve this by moving the screen farther from the projector because you are already using the entire length of the room. Would you change to a lens with a longer or a shorter focal length than the one you are using? Explain your reasoning. (*Hint:* Note that the image distance is essentially fixed.)

63. Mirrors and Lenses Each of the parts of this problem has a description of an object and an optical device (lens or mirror). A sketch is shown in Fig. 35-54. For each case, specify whether

• The image is real (R), virtual (V), or no image is formed (N).
• The image is on the same side of the device as the object (S) or the opposite side (O). If there is no image put a null mark (∅).
• If an image is formed, on

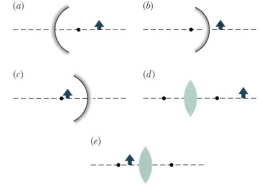

FIGURE 35-54 ■ Problem 63.

which side of the system must the observer be in order to see it, left (−) or right (+)?

For each problem you should therefore give three answers (for example, VO +). For the mirrors, the center is shown. For the lenses, the focal points are shown. The radius of curvature of the mirrors is *R*, and the focal length of the lenses is *f*.

(a) An object on the right side of a spherical mirror, a distance $s > R$ from the mirror. The mirror is concave toward the object.

(b) An object on the right side of a spherical mirror, a distance $s < R/2$ from the mirror. The mirror is convex toward the object.

(c) An object on the left side of a spherical mirror, a distance $R > s > R/2$ from the mirror. The mirror is concave toward the object.

(d) An object on the right side of a convex lens, a distance $s > f$ from the lens.

(e) An object on the left side of a convex lens, a distance $s < f$ from the lens.

(c) An object on the left side of a spherical mirror, a distance $R > s > R/2$ from the mirror. The mirror is concave toward the object.
(d) An object on the right side of a convex lens, a distance $s > f$ from the lens.
(e) An object on the left side of a convex lens, a distance $s < f$ from the lens.

64. The Diverging Lens In Fig. 35-55, point A (marked by a circle) is the top of a small object (indicated as an arrow). Near it is a concave lens, as shown. The focal points of the lens are marked with black dots.

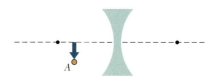

FIGURE 35-55 ■ Problem 64.

(a) Using a ray diagram, show where an image of point A would be formed.
(b) If the focal length of the lens is 8 cm and the object is 6 cm from the lens, where will the image be?
(c) If the object is 1 cm tall, how tall will the image be?
(d) Will the image created by the lens be real or virtual?
(e) Where will you have to be to see the image?

65. The Half Lens A projector has an arrangement of lenses as shown in Fig. 35-56. A bulb illuminates an object (a slide) and the light then passes through a lens that creates an image on a distant screen as shown.

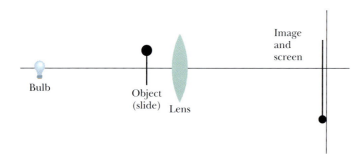

FIGURE 35-56 ■ Problem 65.

When a sheet of a cardboard is brought up to cover the lower half of the lens, what happens to the image on the screen?

(a) The top half of the image disappears.
(b) The bottom half of the image disappears.
(c) The image remains but is weaker (not as bright).
(d) The image remains unchanged.
(e) The bottom half of the image becomes weaker, the top is unchanged.
(f) The top half of the image becomes weaker, the bottom is unchanged.
(g) Something else happens. (Tell what it is.)

Explain your reasoning, drawing whatever rays are needed to make your point clear.

66. Alice and the Looking Glass Alice faces a looking glass (mirror) and is standing at a level so that her eyes appear to her to be right at the top of the mirror as shown in Fig. 35-57. At the position she is standing, she can just see her belt buckle at the bottom of the mirror. If she steps back far enough,

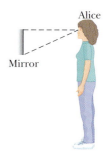

FIGURE 35-57 ■ Problem 66.

(a) She will eventually be able to see all of herself in the mirror at the same time.
(b) There will be no change in how much of herself she can see.
(c) She will see less of herself as she steps back.
(d) Some other result (explain).

Choose the letter of the choice that completes the sentence correctly and explain why you think so with a few sentences and some rays on the diagram.

67. A Bigger Lens Point O in Fig. 35-58 is a source of light. Two rays from O are shown passing through a thin converging lens and crossing each other at the point marked C.

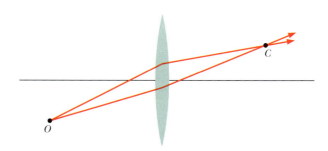

FIGURE 35-58 ■ Problem 67.

(a) On a copy of the figure on your answer sheet, find the two principal foci of the lens by drawing appropriate rays. Label the foci F1 and F2. Explain your reasoning.
(b) Suppose the lens is replaced by another lens having the same focal length but a larger diameter. Indicate whether each of the following partial sentences is correctly completed by the phrase greater than ($>$), less than ($<$), or the same as ($=$).

• The distance of the image from the principal axis is _____ it was with the smaller lens.
• The brightness of the image with the large lens is _____ it was with the smaller lens.

68. Where's the Image? Figure 35-59 shows a thin lens (indicated by a gray rectangle) and a coordinate system. The x axis passes

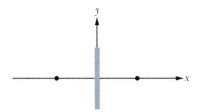

FIGURE 35-59 ■ Problem 68.

through the center of the lens and runs along its axis of symmetry, with the positive x direction indicated by the arrowhead. The lens may be treated as being of negligible thickness and has a focal length f. The points $(x, y) = (f, 0)$ and $(-f, 0)$ are marked by black dots.

A small object is placed at the position $(x, 0)$. For each of the four cases (i)–(iv) below, indicate whether the location of the image formed $(= x')$ is on the positive or negative side of the axis, and closer to the lens than the focal point or farther away.

 i. $-f < x < 0$ iii. $x < f < 0$

 ii. $f < 0 < x$ iv. $0 < f < x$

 Hint: Your answer should take a form such as $(x' > f > 0)$ to indicate that the image is on the positive side of the axis and farther away than the focal point or $(0 > x' > f)$ to indicate that the image is on the negative side of the axis between the lens and the focal point. Note that in some cases the focal length specified is negative and in some cases it is positive.

36 | Interference

At first glance, the top surface of the *Morpho* butterfly's wing is simply a beautiful blue-green. There is something strange about the color, however, for it almost glimmers, unlike the colors of most objects—and if you change your perspective, or if the wing moves, the tint of the color changes. The wing is said to be iridescent, and the blue-green we see hides the wing's "true" dull brown color that appears on the bottom surface.

What is so different about the top surface that gives us this arresting display?

The answer is in this chapter.

36-1 Interference

Sunlight, as the rainbow shows us, is a composite of all the colors of the visible spectrum. The colors reveal themselves in the rainbow because the incident wavelengths refract and so are bent through different angles as they pass through raindrops that produce the bow. However, soap bubbles and oil slicks can also show striking colors produced not by refraction but by constructive and destructive **interference** of light waves. The interfering waves combine either to enhance or to suppress certain colors in the spectrum of the incident sunlight. Interference of light waves is thus a superposition phenomenon. Hence, this chapter is a significant point of connection between much of what we have just learned in Chapters 34 and 35 about electromagnetic waves in general and light in particular and what we learned earlier regarding the interference of waves on a string and sound waves in Chapters 17 and 18.

Interference, which can lead to the selective enhancement or suppression of wavelengths, has many applications. When light encounters an ordinary glass surface, for example, about 4% of the incident energy is reflected, thus weakening the transmitted beam by that amount. This unwanted loss of light can be a real problem in optical systems with many components. A thin, transparent "interference film," deposited on the glass surface, can reduce the amount of reflected light (and thus enhance the transmitted light) by destructive interference. The bluish cast of a camera lens reveals the presence of such a coating. Interference coatings can also be used to enhance—rather than reduce—the ability of a surface to reflect light.

To understand interference, we must go beyond the restrictions of geometrical optics and employ the full power of wave optics. In fact, as you will see, the existence of interference phenomena is perhaps our most convincing evidence that light is a wave—because interference cannot be explained other than with waves.

36-2 Light as a Wave

The first person to advance a convincing wave theory for light was Dutch physicist Christian Huygens, in 1678. Although much less comprehensive than the later electromagnetic theory of Maxwell, Huygens' theory was simpler mathematically and remains useful today. Its great advantages are that it accounts for the laws of reflection and refraction in terms of waves and gives physical meaning to the index of refraction.

Huygens' wave theory is based on a geometrical construction that allows us to tell where a given wavefront will be at any time in the future if we know its present position. This construction is based on **Huygens' principle,** which is:

> All points on a wavefront serve as point sources of spherical secondary wavelets. After a time Δt, the new position of the wavefront will be that of a surface tangent to these secondary wavelets.*

Here is a simple example. At the left in Fig. 36-1, the present location of a wavefront of a plane wave traveling to the right in vacuum is represented by plane ab, perpendicular to the page. Where will the wavefront be at time Δt later? We let several points on plane ab (the dots) serve as sources of spherical secondary wavelets that are emitted at $t = 0$. At time Δt, the radius of all these spherical wavelets will have grown to

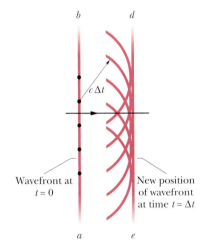

FIGURE 36-1 ■ The propagation of a plane wave in vacuum, as portrayed by Huygens' principle.

---

* When using this principle in calculations, there is a factor which gives a greater wave amplitude in the direction of propagation of the original wavefront. This prevents the back wavelets from combining to create a backwards wavefront.

$c\Delta t$, where c is the speed of light in vacuum. We draw plane *de* tangent to these wavelets at time Δt. This plane represents the wavefront of the plane wave at time Δt; it is parallel to plane *ab* and a perpendicular distance $c\Delta t$ from it.

The Law of Refraction

We now use Huygens' principle to derive the law of refraction or Snell's law, $n_1 \sin\theta_1 = n_2 \sin\theta_2$ (Eq. 35-2). Recall that here n_1 is the index of refraction in the medium from which the light is incident, n_2 is the index of refraction in the refracting medium, and θ_1 and θ_2 are the angles of incidence and refraction, respectively. There are two key ideas behind this derivation:

1. The oscillating electromagnetic wave (with its oscillating *E*- and *B*-fields) hitting the surface of a material drives the electrons in the surface to oscillate and hence to reradiate. As a result, the outgoing wave the electrons produce will have the same frequency as the incoming wave.

2. The wavelength is determined by how far the wave propagates into the media while the electrons at the surface are undergoing one full oscillation. Since the relationship between wavelength λ, period T, and speed of the wave v is $v = \lambda/T$, and the wave speed in a denser medium is slower, the wavelength is smaller in a denser medium.

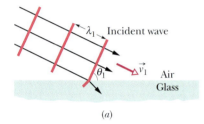

(a)

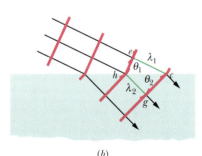

(b)

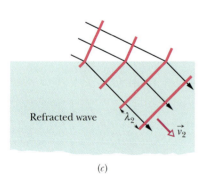

(c)

FIGURE 36-2 ■ The refraction of a plane wave at an air–glass interface, as portrayed by Huygens' principle. The wavelength in glass is smaller than that in air. For simplicity, the reflected wave is not shown. Parts (a) through (c) represent three successive stages of the refraction.

Figure 36-2 shows three stages in the refraction of several wavefronts at a plane interface between air (medium 1) and glass (medium 2). By convention, we choose the wavefronts in the incident beam to be separated by λ_1, the wavelength in medium 1. Let the speed of light in air be v_1 and that in glass be v_2. We assume that $v_2 < v_1$, which happens to be true. (Since in this chapter we do not use vector components, we run no risk of confusing the magnitude of a velocity vector with its components. Therefore, we will simplify notation and write the speed as simply v, rather than $|\vec{v}|$.)

Angle θ_1 in Fig. 36-2a is the angle between the wavefront and the interface; it has the same value as the angle between the *normal* to the wavefront (that is, the incident ray) and the *normal* to the interface. Thus, θ_1 is the angle of incidence.

As the incident light wave moves into the glass, a wavefront at point *e* will travel a distance of λ_1 to point *c*. The time period required for the wave to travel this distance is the distance divided by the speed of the wavelet, or λ_1/v_1. In this same time period, a Huygens wavelet (a new wave created by the oscillating electrons in the material) at point *h* will travel to point *g*, at the reduced speed v_2 and with wavelength λ_2. Thus, this time period must also be equal to λ_2/v_2. By equating these times, we obtain the relation

$$\frac{\lambda_1}{\lambda_2} = \frac{v_1}{v_2}, \qquad (36\text{-}1)$$

which shows that the wavelengths of light in two media are proportional to the speeds of light in those media.

By Huygens' principle, the refracted wavefront must be tangent to an arc of radius λ_2 centered on *h*, say, at point *g*. The refracted wavefront must also be tangent to an arc of radius λ_1 centered on *e*, say, at *c*. Then the refracted wavefront must be oriented as shown. Note that θ_2, the angle between the refracted wavefront and the interface, is actually the angle of refraction.

For the right triangles *hce* and *hcg* in Fig. 36-2b we may write

$$\sin\theta_1 = \frac{\lambda_1}{hc} \qquad \text{(for triangle } hce\text{)}$$

and

$$\sin\theta_2 = \frac{\lambda_2}{hc} \qquad \text{(for triangle } hcg\text{)}.$$

Dividing the first of these two equations by the second and using Eq. 36-1, we find

$$\frac{\sin\theta_1}{\sin\theta_2} = \frac{\lambda_1}{\lambda_2} = \frac{v_1}{v_2}. \tag{36-2}$$

We can define an **index of refraction** n for each medium as the ratio of the speed of light in vacuum to the speed of light v in the medium. Thus,

$$n \equiv \frac{c}{v} \qquad \text{(definition of index of refraction)}. \tag{36-3}$$

In particular, for our two media, we have

$$n_1 = \frac{c}{v_1} \quad \text{and} \quad n_2 = \frac{c}{v_2}. \tag{36-4}$$

If we combine Eqs. 36-2 and 36-4 we find

$$\frac{\sin\theta_1}{\sin\theta_2} = \frac{c/n_1}{c/n_2} = \frac{n_2}{n_1}, \tag{36-5}$$

or

$$n_1 \sin\theta_1 = n_2 \sin\theta_2 \qquad \text{(law of refraction)}, \tag{36-6}$$

as introduced in Chapter 34. This result demonsrates that Huygen's Principle is a construct that can be used to explain the law of refraction.

Wavelength and Index of Refraction

We have now seen that the wavelength of light changes when the speed of the light changes, as happens when light crosses an interface from one medium into another. Further, the speed of light in any medium depends on the index of refraction of the medium, according to Eq. 36-3. Thus, the wavelength of light in any medium depends on the index of refraction of the medium. Let a certain monochromatic light have wavelength λ and speed c in vacuum and wavelength λ_n and speed v in a medium with an index of refraction n. Now we can rewrite Eq. 36-1 as

$$\lambda_n = \lambda\frac{v}{c}. \tag{36-7}$$

Using Eq. 36-3 to substitute $1/n$ for v/c then yields

$$\lambda_n = \frac{\lambda}{n}. \tag{36-8}$$

This equation relates the wavelength of light in any medium to its wavelength in vacuum. It tells us that the greater the index of refraction of a medium, the smaller is the wavelength of light in that medium.

What about the frequency of the light? Let f_n represent the frequency of the light in a medium with index of refraction n. Then from the general relation of Eq. 17-12 ($v = \lambda f$), we can write

$$f_n = \frac{v}{\lambda_n}.$$

Substituting Eqs. 36-3 and 36-8 then gives us

$$f_n = \frac{c/n}{\lambda/n} = \frac{c}{\lambda} = f,$$

where f is the frequency of the light in vacuum. Thus, although the speed and wavelength of light are different in the medium than in vacuum, the frequency of the light in the medium is the same as it is in vacuum. Since we started this discussion with the assumption that the periods of the waves were the same, this result indicates that our equations are consistent.

The fact that the wavelength of light depends on the index of refraction via Eq. 36-8 is important in certain situations involving the interference of light waves. For example, in Fig. 36-3, the *waves of the rays* (that is, the waves associated with the rays) have identical wavelengths λ and are initially in phase in air ($n \approx 1$). One of the waves travels through medium 1 of index of refraction n_1 and length L. The other travels through medium 2 of index of refraction n_2 and the same length L. Each ray acquires a different wavelength when traveling through its medium. When the waves leave the two media, they will have the same wavelength once again—their wavelength λ in air. However, because their wavelengths differed in the two media, the two waves may no longer be in phase.

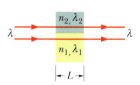

FIGURE 36-3 ■ Two light rays with the same initial wavelength λ in air travel through two media having different indexes of refraction. During that time, $\lambda_1 \neq \lambda_2$.

> The phase difference between two light waves can change if the waves travel through different materials having different indexes of refraction.

As we shall discuss soon, this phase difference change can determine how the light waves will interfere if they reach some common point.

To find their new phase difference in terms of wavelengths, we first count the number N_1 of wavelengths there are in the length L of medium 1. From Eq. 36-8, the wavelength in medium 1 is $\lambda_{n1} = \lambda/n_1$, so

$$N_1 = \frac{L}{\lambda_{n1}} = \frac{Ln_1}{\lambda}. \tag{36-9}$$

In general, the term Ln is known as the **optical path difference**. Similarly, we count the number N_2 of wavelengths there are in the length L of medium 2, where the wavelength is $\lambda_{n2} = \lambda/n_2$:

$$N_2 = \frac{L}{\lambda_{n2}} = \frac{Ln_2}{\lambda}. \tag{36-10}$$

To find the new phase difference between the waves, we subtract the smaller of N_1 and N_2 from the larger. Assuming $n_2 > n_1$, we obtain

$$N_2 - N_1 = \frac{Ln_2}{\lambda} - \frac{Ln_1}{\lambda} = \frac{L}{\lambda}(n_2 - n_1). \tag{36-11}$$

Thus the phase difference is simply the optical path length difference divided by the wavelength of the light in a vacuum.

Suppose Eq. 36-11 tells us that the waves now have a phase difference of 45.6 wavelengths. That is equivalent to taking the initially in-phase waves and shifting one of them by 45.6 wavelengths. However, a shift of an integer number of wavelengths (such as 45) would put the waves back in phase, so it is only the decimal fraction (here, 0.6) that is important. A phase difference of 45.6 wavelengths is equivalent to an *effective phase difference* of 0.6 wavelength.

A phase difference of 0.5 wavelength puts two waves exactly out of phase. If the waves had equal amplitudes and were to reach some common point, they would then undergo fully destructive interference, producing darkness at that point. With an effective phase difference of 0.0 wavelength, they would, instead, undergo fully constructive interference, resulting in brightness at the common point. Our effective phase difference of 0.6 wavelength is an intermediate situation, but closer to destructive interference, and the waves would produce a dimly illuminated common point.

We can also express phase difference in terms of radians and degrees, as we have done already.

> A phase difference of one wavelength is equivalent to phase differences of 2π rad or 360°.

READING EXERCISE 36-1: The figure shows a monochromatic ray of light traveling across parallel interfaces, from an original material a, through layers of material b and c, and then back into material a. Rank the materials according to the speed of light in them, greatest first.

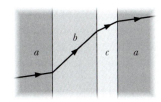

TOUCHSTONE EXAMPLE 36-1: Phase Difference and Interference

In Fig. 36-3, the two light waves that are represented by the rays have wavelength 550.0 nm before entering media 1 and 2. They also have equal amplitudes and are in phase. Medium 1 is now just air, and medium 2 is a transparent plastic layer of index of refraction 1.600 and thickness 2.600 μm.

(a) What is the phase difference of the emerging waves in wavelengths, radians, and degrees? What is their effective phase difference (in wavelengths)?

SOLUTION ■ One **Key Idea** here is that the phase difference of two light waves can change if they travel through different media, with different indexes of refraction. The reason is that their wavelengths are different in the different media. We can calculate the change in phase difference by counting the number of wavelengths that fits into each medium and then subtracting those numbers. When the path lengths of the waves in the two media are identical, Eq. 36-11 gives the result. Here we have $n_1 = 1.000$ (for the air), $n_2 = 1.600$, $L = 2.600$ μm, and $\lambda = 550.0$ nm. Thus, Eq. 36-11 yields

$$N_2 - N_1 = \frac{L}{\lambda}(n_2 - n_1)$$

$$= \frac{2.600 \times 10^{-6}\,\text{m}}{5.500 \times 10^{-7}\,\text{m}}(1.600 - 1.000)$$

$$= 2.84. \qquad \text{(Answer)}$$

Thus, the phase difference of the emerging waves is 2.84 wavelengths. Because 1.0 wavelength is equivalent to 2π rad and 360°, you can show that this phase difference is equivalent to

$$\text{phase difference} = 17.8\,\text{rad} \approx 1020°. \qquad \text{(Answer)}$$

A second **Key Idea** is that the effective phase difference is the decimal part of the actual phase difference *expressed in wavelengths.* Thus, we have

$$\text{effective phase difference} = 0.84\,\text{wavelength.} \qquad \text{(Answer)}$$

You can show that this is equivalent to 5.3 rad and about 300°. *Caution:* We do *not* find the effective phase difference by taking the decimal part of the actual phase difference as expressed in radians or degrees. For example, we do *not* take 0.8 rad from the actual phase difference of 17.8 rad.

(b) If the rays of the waves were angled slightly so that the waves reached the same point on a distant viewing screen, what type of interference would the waves produce at that point?

SOLUTION ■ The **Key Idea** here is to compare the effective phase difference of the waves with the phase differences that give the extreme types of interference. Here the effective phase difference of 0.84 wavelength is between 0.5 wavelength (for fully destructive interference, or the darkest possible result) and 1.0 wavelength (for fully constructive interference, or the brightest possible result), but closer to 1.0 wavelength. Thus, the waves would produce intermediate interference that is closer to fully constructive interference—they would produce a relatively bright spot.

36-3 Diffraction

In the next section we shall discuss the experiment that first proved that light is a wave. To prepare for that discussion, we must introduce the idea of **diffraction** of waves, a phenomenon that we explore much more fully in Chapter 37. Its essence is this: If a wave encounters a barrier that has an opening of dimensions similar to the wavelength, the part of the wave that passes through the opening will flare (spread) out—will *diffract*—into the region beyond the barrier. The flaring out is consistent with the spreading of the wavelets in the Huygens construction of Fig. 36-1. Diffraction occurs for waves of all types, not just light waves. Figure 36-4 shows the diffraction of water waves traveling across the surface of water in a shallow tank.

Figure 36-5a shows the situation schematically for an incident plane wave of wavelength λ encountering a slit that has width $a = 6.0\lambda$ and extends into and out of the page. The wave flares out on the far side of the slit. Figures 36-5b (with $a = 3.0\lambda$) and 36-5c ($a = 1.5\lambda$) illustrate the main feature of diffraction: the narrower the slit, the greater the diffraction.

Diffraction limits geometrical optics, in which we represent an electromagnetic wave with a ray. If we actually try to form a ray by sending light through a narrow slit, or through a series of narrow slits, diffraction will always defeat our effort because it always causes the light to spread. Indeed, the narrower we make the slits (in the hope of producing a narrower beam), the greater the spreading is. Thus, geometrical optics holds only when slits or other apertures that might be located in the path of light have dimensions that are much larger than the wavelength of the light.

FIGURE 36-4 ■ The diffraction of water waves in a ripple tank. The waves are produced by an oscillating paddle at the left. As they move from left to right, they flare out through an opening in a barrier along the water surface.

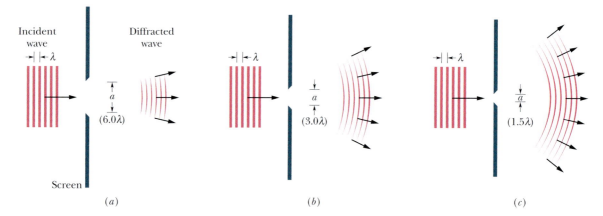

(a) (b) (c)

FIGURE 36-5 ■ Diffraction represented schematically. For a given wavelength λ, the diffraction is more pronounced the smaller the slit width a. The figures show the cases for (a) slit width $a = 6.0\lambda$, (b) slit width $a = 3.0\lambda$, and (c) slit width $a = 1.5\lambda$. In all three cases, the screen and the length of the slit extend well into and out of the page, perpendicular to it.

36-4 Young's Interference Experiment

In 1801, Thomas Young experimentally proved that light is a wave, contrary to what most other scientists then thought. He did so by demonstrating that light undergoes interference, as do water waves, sound waves, and waves of all other types. In addition, he was able to measure the average wavelength of sunlight; his value, 570 nm, is

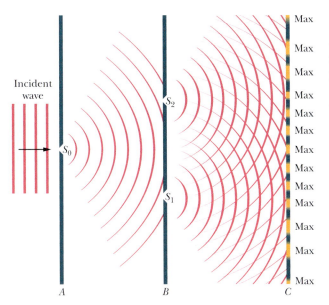

Max
Max
Max
Max
Max
Max
Max
Max
Max
Max
Max
Max
Max

A   *B*   *C*

FIGURE 36-6 ■ In Young's interference experiment, incident monochromatic light is diffracted by slit S_0, which then acts as a point source of light that emits semicircular wavefronts. As that light reaches screen *B*, it is diffracted by slits S_1 and S_2, which then act as two point sources of light. The light waves traveling from slits S_1 and S_2 overlap and undergo interference, forming an interference pattern of maxima and minima on viewing screen *C*. This figure is a cross section; the screens, slits, and interference pattern extend into and out of the page. Between screens *B* and *C*, the semicircular wavefronts centered on S_2 depict the waves that would be there if only S_2 were open. Similarly, those centered on S_1 depict waves that would be there if only S_1 were open.

impressively close to the modern accepted value of 555 nm. We shall here examine Young's historic experiment as an example of the interference of light waves.

Figure 36-6 gives the basic arrangement of Young's experiment. Light from a distant monochromatic source illuminates slit S_0 in screen *A*. The emerging light then spreads via diffraction to illuminate two slits S_1 and S_2 in screen *B*. Diffraction of the light by these two slits sends overlapping circular waves into the region beyond screen *B*, where the waves from one slit interfere with the waves from the other slit.

The "snapshot" of Fig. 36-6 depicts the interference of the overlapping waves from very small slits. However, we cannot see evidence for the interference except where a viewing screen *C* intercepts the light. Where it does so, points of interference maxima form visible bright rows—called *bright bands, bright fringes*, or (loosely speaking) *maxima*—that extend across the screen (into and out of the page in Fig. 36-6). Dark regions—called *dark bands, dark fringes*, or (loosely speaking) *minima*—result from fully destructive interference and are visible between adjacent pairs of bright fringes. (*Maxima* and *minima* more properly refer to the center of a band.) The pattern of bright and dark fringes on the screen is called an **interference pattern.** Figure 36-7 is a photograph of part of the interference pattern as seen from the left in Fig. 36-6. The fringes that appear on a flat screen get further apart and dimmer as the distance from the center of the screen increases.

Note: These are very small slits, so we assume diffraction acts to spread out the waves passing through each slit. This explains why the waves from the two slits can overlap at a distant point. However, for now we ignore the fact that a diffracted wave has more intensity in the direction of the wavefront incident on the slit. In Chapter 37 we will analyze diffraction mathematically, which is the cause of the weakening intensity away from the center of the pattern seen in Fig. 36-7.

Locating the Fringes

Light waves produce fringes in a *Young's double-slit interference experiment*, as it is called, but what exactly determines the locations of the fringes? To answer, we shall use the arrangement in Fig. 36-8a. There, a plane wave of monochromatic light is incident on two slits S_1 and S_2 in screen *B*; the light diffracts through the slits and produces an interference pattern on screen *C*. We draw a central axis from the point halfway between the slits to screen *C* as a reference. We then pick, for discussion, an arbitrary point *P* on the screen, at angle θ to the central axis. This point intercepts the wave of ray r_1 from the bottom slit and the wave of ray r_2 from the top slit.

FIGURE 36-7 ■ A photograph of the interference pattern produced by the arrangement shown in Fig 36-6. (The photograph is a front view of part of screen *C*.) The alternating maxima and minima are called *interference fringes* (because they resemble the decorative fringe sometimes used on clothing and rugs).

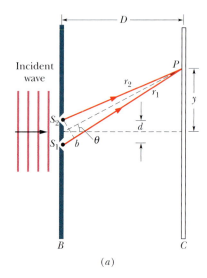

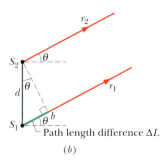

FIGURE 36-8 ■ (a) Waves from slits S_1 and S_2 (which extend into and out of the page) with spacing d combine at P, an arbitrary point on screen C at distance y from the central axis. The angle θ serves as a convenient locator for P. (b) For a slit screen distance of $D \gg d$, we can approximate rays r_1 and r_2 as being parallel, at angle θ to the central axis. *Note:* In a typical demonstration of these effects d might be on the order of 1 mm or some fraction thereof and D might be 1–2 meters or more.

These waves are in phase when they pass through the two slits because there they are just portions of the same incident wave. However, once they have passed the slits, the two waves must travel different distances to reach P. We saw a similar situation in Section 18-4 with sound waves and concluded that

> The phase difference between two waves can change if the waves travel paths of different lengths.

The change in phase difference is due to the *path length difference* ΔL in the paths taken by the waves. Consider two waves initially exactly in phase, traveling along paths with a path length difference ΔL, and then passing through some common point. When ΔL is zero or an integer number of wavelengths, the waves arrive at the common point exactly in phase and they interfere fully constructively there. If that is true for the waves of rays r_1 and r_2 in Fig. 36-8, then point P is part of a bright fringe. When, instead, ΔL is an odd multiple of half a wavelength, the waves arrive at the common point exactly out of phase and they interfere fully destructively there. If that is true for the waves of rays r_1 and r_2, then point P is part of a dark fringe (and, of course, we can have intermediate situations of interference and thus intermediate illumination at P.) Thus,

> What appears at each point on the viewing screen in a Young's double-slit interference experiment is determined by the path length difference ΔL of the rays reaching that point.

We can specify where each bright or dark fringe is located on the screen by giving the angle θ from the central axis to that fringe. To find θ, we must relate it to ΔL. We start with Fig. 36-8a by finding a point b along ray r_1 such that the path length from points b to P equals the path length from S_2 to P. Then the path length difference ΔL between the two rays is the distance from S_1 to b.

The relation between this S_1-to-b distance and θ is complicated, but we can simplify it considerably if we arrange for the distance D from the slits to the screen to be much greater than the slit separation d. Then we can approximate rays r_1 and r_2 as being parallel to each other and at angle θ to the central axis (Fig. 36-8b). We can also approximate the triangle formed by S_1, S_2, and b as being a right triangle, and approximate the angle inside that triangle at S_2 as being θ. Then, for that triangle, $\sin \theta = \Delta L/d$ and thus

$$\Delta L = d \sin\theta \quad \text{(path length difference).} \quad (36\text{-}12)$$

For a bright fringe, we saw that ΔL must be zero or an integer number of wavelengths. Using Eq. 36-12, we can write this requirement as

$$\Delta L = d \sin\theta = (\text{integer})(\lambda), \quad (36\text{-}13)$$

$$\text{or as} \quad d \sin\theta = m\lambda, \quad \text{for } m = 0, 1, 2, \ldots \quad \text{(maxima—bright fringes).} \quad (36\text{-}14)$$

For a dark fringe, ΔL must be an odd multiple of half a wavelength. Again using Eq. 36-12, we can write this requirement as

$$\Delta L = d \sin\theta = (\text{odd number})(\tfrac{1}{2}\lambda), \quad (36\text{-}15)$$

$$\text{or as} \quad d \sin \theta = (m + \tfrac{1}{2})\lambda, \quad \text{for } m = 0, 1, 2, \ldots \quad \text{(minima—dark fringes).} \quad (36\text{-}16)$$

With Eqs. 36-14 and 36-16, we can find the angle θ to any fringe and thus locate that fringe; further, we can use the values of m to label the fringes. For $m = 0$, Eq. 36-14 tells us that a bright fringe is at $\theta = 0$—that is, on the central axis. This *central maximum* is the point at which waves arriving from the two slits have a path length difference $\Delta L = 0$, hence zero phase difference.

For, say, $m = 2$, Eq. 36-14 tells us that *bright* fringes are at

$$\theta = \sin^{-1}\left(\frac{2\lambda}{d}\right)$$

above and below the central axis. Waves from the two slits arrive at these two fringes with $\Delta L = 2\lambda$ and with a phase difference of two wavelengths. These fringes are said to be the *second-order fringes* (meaning $m = 2$) or the *second side maxima* (the second maxima to the side of the central maximum), or they are described as being the second fringes from the central maximum.

For $m = 1$, Eq. 36-16 tells us that *dark* fringes are at

$$\theta = \sin^{-1}\left(\frac{1.5\lambda}{d}\right)$$

above and below the central axis. Waves from the two slits arrive at these two fringes with $\Delta L = 1.5\lambda$ and with a phase difference, in wavelengths, of 1.5. These fringes are called the *second dark fringes* or *second minima* because they are the second dark fringes from the central axis. (The first dark fringes, or first minima, are at locations for which $m = 0$ in Eq. 36-16.)

We derived Eqs. 36-14 and 36-16 for the situation $D \gg d$. However, they also apply if we place a converging lens between the slits and the viewing screen and then move the viewing screen closer to the slits, to the focal point of the lens. (The screen is then said to be in the *focal plane* of the lens; that is, it is in the plane perpendicular to the central axis at the focal point.) One property of a converging lens is that it focuses all rays that are parallel to one another to the same point on its focal plane. Thus, the rays that now arrive at any point on the screen (in the focal plane) were exactly parallel (rather than approximately) when they left the slits. They are like the initially parallel rays in Fig. 35-23a that are directed to a point (the focal point) by a lens.

READING EXERCISE 36-2: In Fig. 36-8, what are ΔL (as a multiple of the wavelength) and the phase difference (in wavelengths) for the two rays if point P is (a) a third side maximum and (b) a third minimum? ■

TOUCHSTONE EXAMPLE 36-2: Distance Between Adjacent Maxima

What is the distance on screen C in Fig. 36-8a between adjacent maxima near the center of the interference pattern? The wavelength λ of the light is 546 nm, the slit separation d is 0.12 mm, and the slit–screen separation D is 55 cm. Assume that θ in Fig. 36-8 is small enough to permit use of the approximations $\sin\theta \approx \tan\theta \approx \theta$, in which θ is expressed in radian measure.

SOLUTION ■ First, let us pick a maximum with a low value of m to ensure that it is near the center of the pattern. Then one **Key Idea** is that, from the geometry of Fig. 36-8a, the maximum's verti-

cal distance y_m from the center of the pattern is related to its angle θ from the central axis by

$$\tan\theta \approx \theta = \frac{y_m}{D}.$$

A second **Key Idea** is that, from Eq. 36-14, this angle θ for the mth maximum is given by

$$\sin\theta \approx \theta = \frac{m\lambda}{d}.$$

If we equate these two expressions for θ and solve for y_m, we find

$$y_m = \frac{m\lambda D}{d}. \qquad (36\text{-}17)$$

For the next farther out maximum, we have

$$y_{m+1} = \frac{(m+1)\lambda D}{d}. \qquad (36\text{-}18)$$

We find the distance between these adjacent maxima by subtracting Eq. 36-17 from Eq. 36-18:

$$\Delta y = y_{m+1} - y_m = \frac{\lambda D}{d}$$

$$= \frac{(546 \times 10^{-9}\,\text{m})(55 \times 10^{-2}\,\text{m})}{0.12 \times 10^{-3}\,\text{m}}$$

$$= 2.50 \times 10^{-3}\,\text{m} \approx 2.5\,\text{mm}. \qquad \text{(Answer)}$$

As long as d and θ in Fig. 36-8a are small, the separation of the interference fringes is independent of m; that is, the fringes are evenly spaced.

36-5 Coherence

For the interference pattern to appear on viewing screen C in Fig. 36-6, the light waves reaching any point P on the screen must have a phase difference that does not vary in time. That is the case in Fig. 36-6, because the waves passing through slits S_1 and S_2 are portions of the single light wave that illuminates the slits. Because the phase difference remains constant, the light from slits S_1 and S_2 is said to be completely **coherent.**

Direct sunlight is partially coherent; that is, sunlight waves intercepted at two points have a constant phase difference only if the points are very close. If you look closely at your fingernail in bright sunlight, you can see a faint interference pattern called *speckle* that causes the nail to appear to be covered with specks. You see this effect because light waves scattering from very close points on the nail are sufficiently coherent to interfere with one another at your eye. The slits in a double-slit experiment, however, are not close enough, and in direct sunlight, the light at the slits would be **incoherent.** To get coherent light, we would have to send the sunlight through a single slit as in Fig. 36-6; because that single slit is small, light that passes through it is coherent. In addition, the smallness of the slit causes the coherent light to spread sufficiently via diffraction to illuminate both slits in the double-slit experiment.

If we replace the double slits with two similar but independent monochromatic light sources, such as two fine incandescent wires, the phase difference between the waves emitted by the sources varies rapidly and randomly. (This occurs because the light is emitted by vast numbers of atoms in the wires, acting randomly and independently for extremely short times—of the order of nanoseconds.) As a result, at any given point on the viewing screen, the interference between the waves from the two sources varies rapidly and randomly between fully constructive and fully destructive. The eye (and most common optical detectors) cannot follow such changes, and no interference pattern can be seen. The fringes disappear, and the screen is seen as being uniformly illuminated.

A *laser* differs from common light sources in that its atoms emit light in a cooperative manner, thereby making the light coherent. Moreover, the light is almost monochromatic, is emitted in a thin beam with little spreading, and can be focused to a width that almost matches the wavelength of the light.

36-6 Intensity in Double-Slit Interference

Equations 36-14 and 36-16 tell us how to locate the maxima and minima of the double-slit interference pattern on screen C of Fig. 36-8 as a function of the angle θ in that figure. Here we wish to derive an expression for the intensity I of the fringes as a function of θ.

The light leaving the slits is in phase. However, let us assume that the light waves from the two slits are not in phase when they arrive at point P. Instead, the electric field components of those waves at point P are not in phase and vary with time as

$$E_1 = E_0 \sin\omega t \tag{36-19}$$

and
$$E_2 = E_0 \sin(\omega t + \phi), \tag{36-20}$$

where ω is the angular frequency of the waves and ϕ is the phase constant of wave E_2. Note that when θ is small the two waves have approximately the same amplitude E_0 and a phase difference of ϕ. Because that phase difference does not vary, the waves are coherent. We shall show that these two waves will combine at P to produce an illumination of intensity I given by

$$I = 4I_0 \cos^2{\tfrac{1}{2}}\phi, \tag{36-21}$$

where

$$\phi = \frac{2\pi d}{\lambda} \sin\theta. \tag{36-22}$$

In Eq. 36-21, I_0 is the intensity of the light that arrives on the screen from one slit when the other slit is temporarily covered. We assume that the slits are so narrow in comparison to the wavelength that this single-slit intensity is essentially uniform over the central region of the screen in which we wish to examine the fringes.

Equations 36-21 and 36-22, which together tell us how the intensity I of the fringe pattern varies with the angle θ in Fig. 36-8, necessarily contain information about the location of the maxima and minima. Let us see if we can extract it.

Study of Eq. 36-21 shows that intensity maxima will occur when

$$\tfrac{1}{2}\phi = m\pi \qquad \text{for } m = 0, 1, 2, \ldots. \tag{36-23}$$

If we put this result into Eq. 36-22, we find

$$2m\pi = \frac{2\pi d}{\lambda} \sin\theta \qquad \text{for } m = 0, 1, 2, \ldots,$$

or
$$\Delta L = d\sin\theta = m\lambda \qquad \text{for } m = 0, 1, 2, \ldots \quad \text{(maxima)}, \tag{36-24}$$

which is exactly Eq. 36-14, the expression that we derived earlier for the locations of the maxima.

The minima in the fringe pattern occur when

$$\tfrac{1}{2}\phi = (m + \tfrac{1}{2})\pi \qquad \text{for } m = 0, 1, 2, \ldots.$$

If we combine this relation with Eq. 36-22 we are led at once to

$$d\sin\theta = (m + \tfrac{1}{2})\lambda \qquad \text{for } m = 0, 1, 2, \ldots \quad \text{(minima)}, \tag{36-25}$$

which is just Eq. 36-16, the expression we derived earlier for the locations of the fringe minima.

Figure 36-9, which is a plot of Eq. 36-21, shows the intensity of double-slit interference pattern fringes near the central maxima as a function of the phase difference ϕ

FIGURE 36-9 ■ A plot of Eq. 36-21, showing the intensity of a double-slit interference pattern as a function of the phase difference between the waves when they arrive from the two slits. I_0 is the (uniform) intensity that would appear on the screen if one slit were covered. The average intensity of the fringe pattern is $2I_0$, and the *maximum* intensity (for coherent light) is $4I_0$.

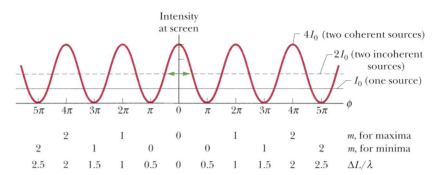

between the waves at the screen. The horizontal solid line is I_0, the (uniform) intensity on the screen when one of the slits is covered up. Note in Eq. 36-21 and the graph that the intensity I (which is always positive) varies from zero at the fringe minima to $4I_0$ at the fringe maxima.

If the waves from the two sources (slits) were *incoherent*, so that no enduring phase relation existed between them, there would be no fringe pattern and the intensity would have the uniform value $2I_0$ for all points on the screen; the horizontal dashed line in Fig. 36-9 shows this uniform value.

Interference cannot create or destroy energy but merely redistributes it over the screen. Thus, the *average* intensity on the screen must be the same $2I_0$ regardless of whether the sources are coherent. This follows at once from Eq. 36-21; if we substitute $\frac{1}{2}$, the average value of the cosine-squared function, this equation reduces to $\langle I \rangle = 2I_0$.

Proof of Eqs. 36-21 and 36-22

We shall combine the electric field components E_1 and E_2, given by Eqs. 36-19 and 36-20, respectively, by the method of phasors discussed in Section 17-12. In Fig. 36-10a, the waves with components E_1 and E_2 are represented by phasors of magnitude E_0 that rotate around the origin at angular speed ω. The values of E_1 and E_2 at any time are the projections of the corresponding phasors on the vertical axis. Figure 36-10a shows the phasors and their projections at an arbitrary time t. Consistent with Eqs. 36-19 and 36-20, the phasor for E_1 has a rotation angle ωt and the phasor for E_2 has a rotation angle $\omega t + \phi$.

To combine the field components E_1 and E_2 at any point P in Fig. 36-8, we add their phasors as if they were vectors (which they are not), as shown in Fig. 36-10b. The magnitude of the phasor sum is the amplitude E of the resultant wave at point P, and that wave has a certain phase constant β. To find the amplitude E in Fig. 36-10b, we first note that the two angles marked β are equal because they are opposite equal-length sides of a triangle. From the theorem (for triangles) that an exterior angle (ϕ) is equal to the sum of the two opposite interior angles ($\beta + \beta$), we see that $\beta = \frac{1}{2}\phi$. Thus, we have

$$E = 2(E_0 \cos \beta)$$
$$= 2E_0 \cos\tfrac{1}{2}\phi. \tag{36-26}$$

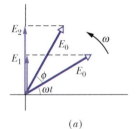

(a)

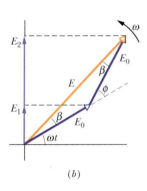

(b)

FIGURE 36-10 ■ (a) Phasors representing, at time t, the electric field components given by Eqs. 36-19 and 36-20. Both phasors have magnitude E_0 and rotate with angular speed ω. Their phase difference is ϕ. (b) Addition of the two phasors gives the phasor representing the resultant wave, with amplitude E and phase constant β.

If we square each side of this relation we obtain

$$E^2 = 4E_0^2 \cos^2 \tfrac{1}{2}\phi. \tag{36-27}$$

Now, from Eq. 34-24, we know that the intensity of an electromagnetic wave is proportional to the square of its amplitude. Therefore, the waves we are combining in Fig. 36-10b, whose amplitudes are E_0, each have an intensity I_0 that is proportional to

E_0^2, and the resultant wave, with amplitude E, has an intensity I that is proportional to E^2. Thus,

$$\frac{I}{I_0} = \frac{E^2}{E_0^2}.$$

Substituting Eq. 36-27 into this equation and rearranging then yield

$$I = 4I_0 \cos^2 \tfrac{1}{2}\phi,$$

which is Eq. 36-21, which we set out to prove.

It remains to prove Eq. 36-22, which relates the phase difference ϕ between the waves arriving at any point P on the screen of Fig. 36-8 to the angle θ that serves as a locator of that point.

The phase difference ϕ in Eq. 36-20 is associated with the path difference $\Delta L = S_1 b$ in Fig. 36-8b. If ΔL is $\tfrac{1}{2}\lambda$, then ϕ is π; if ΔL is λ, then ϕ is 2π, and so on. This suggests

$$\text{(phase difference)} = \frac{2\pi}{\lambda} \text{ (path length difference)}. \qquad (36\text{-}28)$$

The path difference ΔL in Fig. 36-8b is $d \sin \theta$, so Eq. 36-28 becomes

$$\phi = \frac{2\pi d}{\lambda} \sin\theta,$$

which is Eq. 36-22, the other equation that we set out to prove.

Combining More Than Two Waves

In a more general case, we might want to find the resultant of more than two sinusoidally varying waves at a point. The general procedure is this:

1. Construct a series of phasors representing the waves to be combined. Draw them end to end, maintaining the proper phase relations between adjacent phasors.

2. Construct the vector-like phasor sum of this array. The length of this sum gives the amplitude of the resultant phasor. The angle between the phasor sum and the first phasor is the phase of the resultant with respect to this first phasor. The projection of this resultant-sum phasor on the vertical axis gives the time variation of the resultant wave.

TOUCHSTONE EXAMPLE 36-3: Three Light Waves

Three light waves combine at a certain point where their electric field components are

$$E_1 = E_0 \sin \omega t,$$
$$E_2 = E_0 \sin(\omega t + 60°),$$
$$E_3 = E_0 \sin(\omega t - 30°).$$

Find their resultant component $E(t)$ at that point.

SOLUTION ■ The resultant wave is

$$E(t) = E_1(t) + E_2(t) + E_3(t).$$

The **Key Idea** here is two-fold: We can use the method of phasors to find this sum, and we are free to evaluate the phasors at any time t. To simplify the solution we choose $t = 0$, for which the phasors representing the three waves are shown in Fig. 36-11. We can add these three phasors either directly on a vector-capable calcula-

FIGURE 36-11 ■ Three phasors, representing waves with equal amplitudes E_0 and with phase constants $0°$, $60°$, and $230°$, shown at time $t = 0$. The phasors combine to give a resultant phasor with magnitude E_R, at angle β.

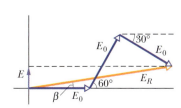

tor or by components. For the component approach, we first write the sum of their horizontal components as

$$\sum E_h = E_0 \cos 0 + E_0 \cos 60° + E_0 \cos(-30°) = 2.37E_0.$$

The sum of their vertical components, which is the value of E at $t = 0$, is

$$\sum E_v = E_0 \sin 0 + E_0 \sin 60° + E_0 \sin(-30°) = 0.366E_0.$$

The resultant wave $E(t)$ thus has an amplitude E_R of

$$E_R = \sqrt{(2.37E_0)^2 + (0.366E_0)^2} = 2.4E_0,$$

and a phase angle β relative to the phasor representing E_1 of

$$\beta = \tan^{-1}\left(\frac{0.366E_0}{2.37E_0}\right) = 8.8°.$$

We can now write, for the resultant wave $E(t)$,

$$E = E_R \sin(\omega t + \beta)$$
$$= 2.4\, E_0 \sin(\omega t + 8.8°). \qquad \text{(Answer)}$$

Be careful to interpret the angle β correctly in Fig. 36-11: It is the constant angle between E_R and the phasor representing E_1 as the four phasors rotate as a single unit around the origin. The angle between E_R and the horizontal axis in Fig. 36-11 does not remain equal to β.

36-7 Interference from Thin Films

The colors we see when sunlight illuminates a soap bubble or an oil slick are caused by the interference of light waves reflected from the front and back surfaces of a thin transparent film. The thickness of the soap or oil film is typically of the order of magnitude of the wavelength of the (visible) light involved. (We shall not consider greater thicknesses, which spoil the coherence of the light needed to produce colors by interference; we shall discuss lesser thicknesses shortly.)

Figure 36-12 shows a thin transparent film of uniform thickness L and index of refraction n_2, illuminated by bright light of wavelength λ from a distant point source. For now, we assume that air lies on both sides of the film and thus that $n_1 = n_3$ in Fig. 36-12. For simplicity, we also assume that the light rays are almost perpendicular to the film ($\theta \approx 0$). We are interested in whether the film is bright or dark to an observer viewing it almost perpendicularly. (Since the film is brightly illuminated, how could it possibly be dark? You will see.)

The incident light, represented by ray i, hits the front (left) surface of the film at point a and undergoes both reflection and refraction there. The reflected ray r_1 enters the observer's eye. The refracted light crosses the film to point b on the back surface, where it undergoes both reflection and refraction. The light reflected at b crosses back through the film to point c, where it undergoes both reflection and refraction. The light refracted at c, represented by ray r_2, also enters the observer's eye.

If the light waves of rays r_1 and r_2 are exactly in phase at the eye, they produce an interference maximum, and region ac on the film is bright to the observer. If they are exactly out of phase, they produce an interference minimum, and region ac is dark to the observer, *even though it is illuminated.* If there is some intermediate phase difference, there are intermediate interference and intermediate brightness.

Thus, the key to what the observer sees is the phase difference between the waves of rays r_1 and r_2. Both rays are derived from the same ray i, but the path involved in producing r_2 involves light traveling twice across the film (a to b, and then b to c), whereas the path involved in producing r_1 involves no travel through the film. Because θ is close to zero, we approximate the path length difference between the waves of r_1 and r_2 as $2L$. However, to find the phase difference between the waves, we cannot just find the number of wavelengths λ that is equivalent to a path length difference of $2L$. This simple approach is impossible for two reasons: (1) the path length difference occurs in a medium other than air, and (2) the reflections involved can change the phase.

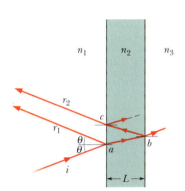

FIGURE 36-12 ■ Light waves, represented with ray i, are incident on a thin film of thickness L and index of refraction n_2. Rays r_1 and r_2 represent light waves that have been reflected by the front and back surfaces of the film, respectively. (All three rays are actually nearly perpendicular to the film.) The interference of the waves of r_1 and r_2 with each other depends on their phase difference. The index of refraction n_1 of the medium at the left can differ from the index of refraction n_3 of the medium at the right, but for now we assume that both media are air, with $n_1 = n_3 = 1.0$, which is less than n_2.

The phase difference between two waves can change if one or both are reflected.

Before we continue our discussion of interference from thin films, we must discuss changes in phase that are caused by reflections.

Reflection Phase Shifts

Refraction at an interface never causes a phase change—but reflection can, depending on the indices of refraction on the two sides of the interface. Figure 36-13 shows what happens when reflection of light waves causes a phase change, using as an example mechanical wave pulses on a denser string (along which pulse travel is relatively slow) and a lighter string (along which pulse travel is relatively fast). This effect is just like the one we discussed in Section 17-10. Recall that a pulse traveling down a string will reflect differently from an end of the string that is tied to a post (a fixed end) than from an end that is tied to a ring that can slide on the post (an open end) (see Fig. 17-25). Reflecting off a heavy string is like reflecting off a fixed end. Reflecting off a light string is like reflecting off an open end.

When a pulse traveling relatively slowly along the denser string in Fig. 36-13a reaches the interface with the lighter string, the pulse is partially transmitted and partially reflected, with no change in orientation. For light, this situation corresponds to the incident wave traveling in the medium of greater index of refraction n (recall that greater n means slower speed). In that case, the wave that is reflected at the interface does not undergo a change in phase; that is, the *reflection phase shift* is zero.

When a pulse traveling more quickly along the lighter string in Fig. 36-13b reaches the interface with the denser string, the pulse is again partially transmitted and partially reflected. The transmitted pulse again has the same orientation as the incident pulse, but now the reflected pulse is inverted. For a sinusoidal wave, such an inversion involves a phase change of π rad, or half a wavelength. For light, this situation corresponds to the incident wave traveling in the medium of lesser index of refraction (with greater speed). In that case, the wave that is reflected at the interface undergoes a phase shift of π rad, or half a wavelength.

We can summarize these results for light in terms of the index of refraction of the medium off which (or from which) the light reflects:

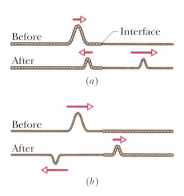

FIGURE 36-13 ■ Phase changes when a pulse is reflected at the interface between two stretched strings of different linear densities. The wave speed is greater in the lighter string. (a) The incident pulse is in the denser string. (b) The incident pulse is in the lighter string. Only here is there a phase change, and only in the reflected wave.

Reflection	Reflection phase shift	Phase change (rad)
Off lower index	$0.0\ \lambda$	0
Off higher index	$0.5\ \lambda$	π

An aid to remembering this is "reflection off high, phase change is pi."

Equations for Thin-Film Interference

In this chapter we have now seen three ways in which the phase difference between two light waves can change:

1. by reflection,

2. by the waves traveling along paths of different lengths,

3. by the waves traveling through media of different indexes of refraction.

When light reflects from a thin film, producing the waves of rays r_1 and r_2 in Fig. 36-12, all three ways are involved. Let us consider them one by one.

We first reexamine the two reflections in Fig. 36-12. At point a on the front interface, the incident wave (in air) reflects from the medium having the higher of the two indexes of refraction, so the wave of reflected ray r_1 has its phase shifted by 0.5 wavelength. At point b on the back interface, the incident wave reflects from the medium

TABLE 36-1
An Organizing Table for Thin-Film Interference in Air[a]

Reflection phase shifts	r_1	r_2
	0.5	0
	wavelength	
Path length difference		$2L$
Index in which path length difference occurs		n_2

[a]Valid for $n_2 > n_1$ and $n_2 > n_3$

(air) having the lower of the two indexes of refraction, so the wave reflected there is not shifted in phase by the reflection, and thus neither is the portion of it that exits the film as ray r_2. We can organize this information with the first line in Table 36-1. It tells us that, so far, as a result of the reflection phase shifts, the waves of r_1 and r_2 have a phase difference of 0.5 wavelength and thus are exactly out of phase.

Based on the information presented in Table 36-1, we have the following rules for predicting constructive interference (in phase) and destructive interference (out of phase):

For a film thickness L, two waves of wavelength λ traveling through a film with index of refraction n_2 will:

constructively interfere (in phase) if $\quad 2L = \dfrac{\text{odd number}}{2} \times \dfrac{\lambda}{n_2}$,

destructively interfere (out of phase) if $\quad 2L = \text{integer} \times \dfrac{\lambda}{n_2}$.

These equations are valid for $n_2 > n_1$ and $n_2 > n_3$.

Now we must consider the path length difference $2L$ that occurs because the wave of ray r_2 crosses the film twice. (This difference $2L$ is shown in the expressions above. If the waves of r_1 and r_2 are to be exactly in phase so that they produce fully constructive interference, the path length $2L$ must cause an additional phase difference of 0.5, 1.5, 2.5, . . . wavelengths. Only then will the net phase difference be an integer number of wavelengths. Thus, for a bright film, we must have

$$2L = \frac{\text{odd number}}{2} \times \text{wavelength} \qquad \text{(in-phase waves).} \qquad (36\text{-}29)$$

The wavelength we need here is the wavelength λ_{n2} of the light in the medium containing path length $2L$—that is, in the medium with index of refraction n_2. Thus, we can rewrite Eq. 36-29 as

$$2L = \frac{\text{odd number}}{2} \times \lambda_{n2} \qquad \text{(in-phase waves).} \qquad (36\text{-}30)$$

If, instead, the waves are to be exactly out of phase so that there is fully destructive interference, the path length $2L$ must cause either no additional phase difference or a phase difference of 1, 2, 3, . . . , wavelengths. Only then will the net phase difference be an odd number of half-wavelengths. For a dark film, we must have

$$2L = \text{integer} \times \text{wavelength}, \qquad (36\text{-}31)$$

where, again, the wavelength is the wavelength $\lambda_{n\,2}$ in the medium containing $2L$. Thus, this time we have

$$2L = \text{integer} \times \lambda_{n\,2} \qquad \text{(out-of-phase waves).} \qquad (36\text{-}32)$$

Now we can use Eq. 36-8 ($\lambda_n = \lambda/n$) to write the wavelength of the wave of ray r_2 inside the film as

$$\lambda_{n\,2} = \frac{\lambda}{n_2}, \qquad (36\text{-}33)$$

where λ is the wavelength of the incident light in vacuum (and approximately also in air). Substituting Eq. 36-33 into Eq. 36-30 and replacing "odd number/2" with $(m + \tfrac{1}{2})$ give us

$$2L = (m + \tfrac{1}{2})\frac{\lambda}{n_2} \qquad \text{for } m = 0, 1, 2, \ldots \qquad \text{(maxima—bright film in air).} \qquad (36\text{-}34)$$

Similarly, with m replacing "integer," Eq. 36-32 yields

$$2L = m\frac{\lambda}{n_2} \qquad \text{for } m = 0, 1, 2, \ldots \qquad \text{(minima—dark film in air).} \qquad (36\text{-}35)$$

For a given film thickness L, Eqs. 36-34 and 36-35 tell us the wavelengths of light for which the film appears bright and dark, respectively, one wavelength for each value of m. Intermediate wavelengths give intermediate brightnesses. For a given wavelength λ, Eqs. 36-34 and 36-35 tell us the thicknesses of the films that appear bright and dark in that light, respectively, one thickness for each value of m. Intermediate thicknesses give intermediate brightnesses.

A special situation arises when a film is so thin that L is much less than λ, say, $L < 0.1\lambda$. Then the path length difference $2L$ can be neglected, and the phase difference between r_1 and r_2 is due *only* to reflection phase shifts. If the film of Fig. 36-12, where the reflections cause a phase difference of 0.5 wavelength, has thickness $L < 0.1\lambda$, then r_1 and r_2 are exactly out of phase, and thus the film is dark, regardless of the wavelength and even the intensity of the light that illuminates it. This special situation corresponds to $m = 0$ in Eq. 36-35. We shall count any thickness $L < 0.1\lambda$ as being the least thickness specified by Eq. 36-35 to make the film of Fig. 36-12 dark. (Every such thickness will correspond to $m = 0$.) The next greater thickness that will make the film dark is that corresponding to $m = 1$.

Figure 36-14 shows a vertical soap film whose thickness increases from top to bottom because the weight of the film has caused it to slump. Bright white light illuminates the film. However, the top portion is so thin that it is dark. In the (somewhat thicker) middle we see fringes, or bands, whose color depends primarily on the wavelength at which reflected light undergoes fully constructive interference for a particular thickness. Toward the (thickest) bottom of the film the fringes become progressively narrower and the colors begin to overlap and fade.

Iridescence of a *Morpho* Butterfly Wing

A surface that displays colors due to thin-film interference is said to be *iridescent* because the tints of the colors change as you change your view of the surface. The iridescence of the top surface of a *Morpho* butterfly wing is due to thin-film interference of light reflected by thin terraces of transparent cuticle-like material on the wing. These terraces are arranged like wide, flat branches on a tree-like structure that extends perpendicular to the wing.

FIGURE 36-14 ◼ The reflection of light from a soapy water film spanning a vertical loop. The top portion is so thin that the light reflected there undergoes destructive interference, making that portion dark. Colored interference fringes, or bands, decorate the rest of the film but are marred by circulation of liquid within the film as the liquid is gradually pulled downward by gravitation.

Suppose you look directly down on these terraces as white light shines directly down on the wing. Then the light reflected back up to you from the terraces undergoes fully constructive interference in the blue-green region of the visible spectrum. Light in the yellow and red regions, at the opposite end of the spectrum, is weaker because it undergoes only intermediate interference. Thus, the top surface of the wing looks blue-green to you.

If you intercept light that reflects from the wing in some other direction, the light has traveled along a slanted path through the terraces. Then the wavelength at which there is fully constructive interference is somewhat different from that for light reflected directly upward. Thus, if the wing moves in your view so that the angle at which you view it changes, the color at which the wing is brightest changes somewhat, producing the iridescence or brilliant rainbow-like colors of the wing.

TOUCHSTONE EXAMPLE 36-4: Brightest Reflected Light

White light, with a uniform intensity across the visible wavelength range of 400 to 690 nm, is perpendicularly incident on a water film, of index of refraction $n_2 = 1.33$ and thickness $L = 320$ nm, that is suspended in air. At what wavelength λ is the light reflected by the film brightest to an observer?

SOLUTION ■ The **Key Idea** here is that the reflected light from the film is brightest at the wavelengths λ for which the reflected rays are in phase with one another. The equation relating these wavelengths λ to the given film thickness L and film index of refraction n_2 is either Eq. 36-34 or Eq. 36-35, depending on the reflection phase shifts for this particular film.

To determine which equation is needed, we should fill out an organizing table like Table 36-1. However, because there is air on both sides of the water film, the situation here is exactly like that in Fig. 36-12, and thus the table would be exactly like Table 36-1. Then from Table 36-1, we see that the reflected rays are in phase (and thus the film is brightest) when

$$2L = \frac{\text{odd number}}{2} \times \frac{\lambda}{n_2},$$

which leads to Eq. 36-34:

$$2L = (m + \tfrac{1}{2})\frac{\lambda}{n_2}.$$

Solving for λ and substituting for L and n_2, we find

$$\lambda = \frac{2n_2 L}{m + \frac{1}{2}} = \frac{(2)(1.33)(320 \text{ nm})}{m + \frac{1}{2}} = \frac{851 \text{ nm}}{m + \frac{1}{2}}.$$

For $m = 0$, this gives us $\lambda = 1700$ nm, which is in the infrared region. For $m = 1$, we find $\lambda = 567$ nm, which is yellow-green light, near the middle of the visible spectrum. For $m = 2$, $\lambda = 340$ nm, which is in the ultraviolet region. Thus, the wavelength at which the light seen by the observer is brightest is

$$\lambda = 567 \text{ nm}. \qquad \text{(Answer)}$$

TOUCHSTONE EXAMPLE 36-5: Magnesium Fluoride Film

In Fig. 36-15, a glass lens is coated on one side with a thin film of magnesium fluoride (MgF_2) to reduce reflection from the lens surface. The index of refraction of MgF_2 is 1.38; that of the glass is 1.50. What is the least coating thickness that eliminates (via interference) the reflections at the middle of the visible spectrum ($\lambda = 550$ nm)? Assume that the light is approximately perpendicular to the lens surface.

SOLUTION ■ The **Key Idea** here is that reflection is eliminated if the film thickness L is such that light waves reflected from the two film interfaces are exactly out of phase. The equation relating L to the given wavelength λ and the index of refraction n_2 of the thin film is either Eq. 36-34 or Eq. 36-35, depending on the reflection phase shifts at the interfaces.

FIGURE 36-15 ■ Unwanted reflections from glass can be suppressed (at a chosen wavelength) by coating the glass with a thin transparent film of magnesium fluoride of the properly chosen thickness.

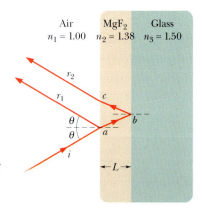

To determine which equation is needed, we fill out an organizing table like Table 36-1. At the first interface, the incident light is in air, which has a lesser index of refraction than the MgF_2 (the thin film). Thus, we fill in 0.5 wavelength under r_1 in our organizing table (meaning that the waves of ray r_1 are shifted by 0.5λ at the first interface). At the second interface, the incident light is in the MgF_2, which has a lesser index of refraction than the glass on the other side of the interface. Thus, we fill in 0.5 wavelength under r_2 in our table.

Because both reflections cause the same phase shift, they tend to put the waves of r_1 and r_2 in phase. Since we want those waves to be *out of phase*, their path length difference $2L$ must be an odd number of half-wavelengths:

$$2L = \frac{\text{odd number}}{2} \times \frac{\lambda}{n_2}.$$

This leads to Eq. 36-34. Solving that equation for L then gives us the film thicknesses that will eliminate reflection from the lens and coating:

$$L = (m + \tfrac{1}{2})\frac{\lambda}{2n_2}, \qquad \text{for } m = 0, 1, 2, \ldots . \quad (36\text{-}36)$$

We want the least thickness for the coating—that is, the least L. Thus, we choose $m = 0$, the least possible value of m. Substituting it and the given data in Eq. 36-36, we obtain

$$L = \frac{\lambda}{4n_2} = \frac{550 \text{ nm}}{(4)(1.38)} = 99.6 \text{ nm.} \qquad \text{(Answer)}$$

TOUCHSTONE EXAMPLE 36-6: Red Light

Figure 36-16a shows a transparent plastic block with a thin wedge of air at the right. (The wedge thickness is exaggerated in the figure.) A broad beam of red light, with wavelength $\lambda = 632.8$ nm, is directed downward through the top of the block (at an incidence angle of 0°). Some of the light is reflected back up from the top and bottom surfaces of the wedge, which acts as a thin film (of air) with a thickness that varies uniformly and gradually from L_L at the left-hand end to L_R at the right-hand end. (The plastic layers above and below the wedge of air are too thick to act as thin films.) An observer looking down on the block sees an interference pattern consisting of six dark fringes and five bright red fringes along the wedge. What is the change in thickness $\Delta L = (L_R - L_L)$ along the wedge?

SOLUTION ■ One **Key Idea** here is that the brightness at any point along the left–right length of the air wedge is due to the interference of the waves reflected at the top and bottom interfaces of the wedge. A second **Key Idea** is that the variation of brightness in the pattern of bright and dark fringes is due to the variation in the thickness of the wedge. In some regions, the thickness puts the reflected waves in phase and thus produces a bright reflection (a bright red fringe). In other regions, the thickness puts the reflected waves out of phase and thus produces no reflection (a dark fringe).

Because the observer sees more dark fringes than bright fringes, we can assume that a dark fringe is produced at both the left and right ends of the wedge. Thus, the interference pattern is that shown in Fig. 36-16b, which we can use to determine the change in thickness ΔL of the wedge.

Another **Key Idea** is that we can represent the reflection of light at the top and bottom interfaces of the wedge, at any point along its length, with Fig. 36-16c, in which L is the wedge thickness at that point. Let us apply this figure to the left end of the wedge, where the reflections give a dark fringe.

We know that, for a dark fringe, the waves of rays r_1 and r_2 in Fig. 36-16c must be out of phase. We also know that the equation relating the film thickness L to the light's wavelength λ and the film's index of refraction n_2 is either Eq. 36-34 or Eq. 36-35, depending on the reflection phase shifts. To determine which equation gives a dark fringe at the left end of the wedge, we should fill out an organizing table like Table 36-1.

At the top interface of the wedge, the incident light is in the plastic, which has a greater index of refraction than the air beneath that interface. Thus, we fill in 0 under r_1 in our organizing table. At the bottom interface of the wedge, the incident light is in air, which has a lesser index of refraction than the plastic beneath that interface. Thus, we fill in 0.5 wavelength under r_2 in our organizing table. Therefore, the reflections alone tend to put the waves of r_1 and r_2 out of phase.

Since the waves are, in fact, out of phase at the left end of the air wedge, the path length difference $2L$ at that end of the wedge must be given by

$$2L = \text{integer} \times \frac{\lambda}{n_2},$$

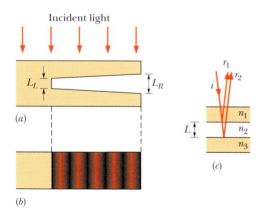

(a)

(b)

(c)

FIGURE 36-16 ■ (a) Red light is incident on a thin, air-filled wedge in the side of a transparent plastic block. The thickness of the wedge is L_L at the left end and L_R at the right end. (b) The view from above the block: an interference pattern of six dark fringes and five bright red fringes lies over the region of the wedge. (c) A representation of the incident ray i, reflected rays r_1 and r_2, and thickness L of the wedge anywhere along the length of the wedge.

which leads to Eq. 36-35:

$$2L = m\frac{\lambda}{n_2}, \qquad \text{for } m = 0, 1, 2, \ldots. \qquad (36\text{-}37)$$

Here is another **Key Idea:** Eq. 36-37 holds not only for the left end of the wedge but also at any point along the wedge where a dark fringe is observed, including the right end—with a different integer value of m for each fringe. The least value of m is associated with the least thickness of the wedge where a dark fringe is observed. Progressively greater values of m are associated with progressively greater thicknesses of the wedge where a dark fringe is observed. Let m_L be the value at the left end. Then the value at the right end must be $m_L + 5$ because, from Fig. 36-16b, the right end is located at the fifth dark fringe from the left end.

We want the change ΔL in thickness, from the left end to the right end of the wedge. To find it we first solve Eq. 36-37 twice—once for the thickness L_L at the left end and once for the thickness L_R at the right end:

$$L_L = (m_L)\frac{\lambda}{2n_2}, \qquad L_R = (m_L + 5)\frac{\lambda}{2n_2}. \qquad (36\text{-}38)$$

To find the change in thickness ΔL, we can now subtract L_L from L_R and substitute known data, including $n_2 = 1.00$ for the air within the wedge:

$$\Delta L = L_R - L_L = \frac{(m_L + 5)\lambda}{2n_2} - \frac{m_L\lambda}{2n_2} = \frac{5}{2}\frac{\lambda}{n_2}$$

$$= \frac{5}{2}\frac{632.8 \times 10^{-9}\text{ m}}{1.00}$$

$$= 1.58 \times 10^{-6}\text{ m}. \qquad \text{(Answer)}$$

36-8 Michelson's Interferometer

An **interferometer** is a device that can be used to measure lengths or changes in length with great accuracy by means of interference fringes. We describe the form originally devised and built by A. A. Michelson in 1881.

Consider light that leaves point P on extended source S in Fig. 36-17 and encounters *beam splitter M*. A beam splitter is a mirror that transmits half the incident light and reflects the other half. In the figure we have assumed, for convenience, that this mirror possesses negligible thickness. At M the light thus divides into two waves. One proceeds by transmission toward mirror M_1; the other proceeds by reflection toward mirror M_2. The waves are entirely reflected at these mirrors and are sent back along their directions of incidence, each wave eventually entering telescope T. What the observer sees is a pattern of curved or approximately straight interference fringes; in the latter case the fringes resemble the stripes on a zebra.

The path length difference for the two waves when they recombine at the telescope is $2d_2 - 2d_1$, and anything that changes this path length difference will cause a change in the phase difference between these two waves at the eye. As an example, if mirror M_2 is moved by a distance $\frac{1}{2}\lambda$, the path length difference is changed by λ and the fringe pattern is shifted by one fringe (as if each dark stripe on a zebra had moved to where the adjacent dark stripe had been). Similarly, moving mirror M_2 by $\frac{1}{4}\lambda$ causes a shift by half a fringe (each dark zebra stripe shifts to where the adjacent white stripe was).

A shift in the fringe pattern can also be caused by the insertion of a thin transparent material into the optical path of one of the mirrors—say, M_1. If the material has thickness L and index of refraction n, then the number of wavelengths along the light's to-and-fro path through the material is, from Eq. 36-9,

$$N_m = \frac{2L}{\lambda_n} = \frac{2Ln}{\lambda}. \qquad (36\text{-}39)$$

The number of wavelengths in the same thickness $2L$ of air before the insertion of the material is

$$N_a = \frac{2L}{\lambda}. \qquad (36\text{-}40)$$

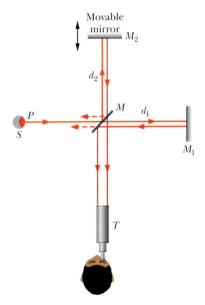

FIGURE 36-17 ■ Michelson's interferometer, showing the path of light originating at point P of an extended source S. Mirror M splits the light into two beams, which reflect from mirrors M_1 and M_2 back to M and then to telescope T. In the telescope an observer sees a pattern of interference fringes.

When the material is inserted, the light returned by mirror M_1 undergoes a phase change (in terms of wavelengths) of

$$N_m - N_a = \frac{2Ln}{\lambda} - \frac{2L}{\lambda} = \frac{2L}{\lambda}(n-1). \qquad (36\text{-}41)$$

For each phase change of one wavelength, the fringe pattern is shifted by one fringe. Thus, by counting the number of fringes through which the material causes the pattern to shift, and substituting that number for $N_m - N_a$ in Eq. 36-41, you can determine the thickness L of the material in terms of λ.

By such techniques the lengths of objects can be expressed in terms of the wavelengths of light. In Michelson's day, the standard of length—the meter—was chosen by international agreement to be the distance between two fine scratches on a certain metal bar preserved at Sèvres, near Paris. Michelson was able to show, using his interferometer, that the standard meter was equivalent to 1 553 163.5 wavelengths of a certain monochromatic red light emitted from a light source containing cadmium. For this careful measurement, Michelson received the 1907 Nobel Prize in physics. His work laid the foundation for the eventual abandonment (in 1961) of the meter bar as a standard of length and for the redefinition of the meter in terms of the wavelength of light. By 1983, as we have seen, even this wavelength standard was not precise enough to meet the growing requirements of science and technology, and it was replaced with a new standard based on a defined value for the speed of light as discussed in Section 1-6.

Problems

SEC. 36-2 ■ LIGHT AS A WAVE

1. Yellow Sodium Light The wavelength of yellow sodium light in air is 589 nm. (a) What is its frequency? (b) What is its wavelength in glass whose index of refraction is 1.52? (c) From the results of (a) and (b) find its speed in this glass.

2. Sapphire vs. Diamond How much faster, in meters per second, does light travel in sapphire than in diamond? See Table 35-1.

3. Yellow Light The speed of yellow light (from a sodium lamp) in a certain liquid is measured to be 1.92×10^8 m/s. What is the index of refraction of this liquid for the light at this wavelength?

4. Fused Quartz What is the speed in fused quartz of light of wavelength 550 nm? (See Fig. 35-3.)

5. Ocean Wave Ocean waves moving at a speed of 4.0 m/s are approaching a beach at an angle of 30° to the normal, as shown from above in Fig. 36-18. Suppose the water depth changes abruptly at a certain distance from the beach and the wave speed there drops to 3.0 m/s. Close to the beach, what is the angle θ between the direction of wave motion and the normal? (Assume the same law of refraction as for light.) Explain why most waves come in normal to a shore even though at large distances they approach at a variety of angles.

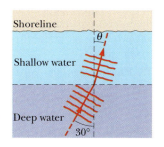

FIGURE 36-18 ■

Problem 5.

6. Two Pulses In Fig. 36-19. Two pulses of light are sent through layers of plastic with the indexes of refraction indicated and with thicknesses of either L or $2L$ as shown. (a) Which pulse travels through the plastic in less time? (b) In terms of L/c, what is the difference in the traversal times of the pulses?

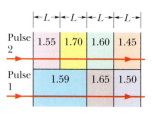

FIGURE 36-19 ■

Problem 6.

7. Two Waves In Fig. 36-3, assume that two waves of light in air, of wave length 400 nm, are initially in phase. One travels through a glass layer of index of refraction $n_1 = 1.60$ and thickness L. The other travels through an equally thick plastic layer of index of refraction $n_2 = 1.50$. (a) What is the least value L should have if the waves are to end up with a phase difference of 5.65 rad? (b) If the waves arrive at some common point after emerging, what type of interference do they undergo?

8. Two Media Suppose that the two waves in Fig. 36-3 have wavelength 500 nm in air. In wavelengths, what is their phase difference after traversing media 1 and 2 if (a) $n_1 = 1.50$, $n_2 = 1.60$, and $L = 8.50$ μm; (b) $n_1 = 1.62$, $n_2 = 1.72$, and $L = 8.50$ μm; and (c) $n_1 = 1.59$, $n_2 = 1.79$, and $L = 3.25$ μm? (d) Suppose that in each of these three situations the waves arrive at a common point after emerging. Rank the situations according to the brightness the waves produce at the common point.

9. Initially in Phase Two waves of light in air, of wavelength 600.0 nm, are initially in phase. They then travel through plastic layers as shown in Fig. 36-20, with $L_1 = 4.00$ μm, $L_2 = 3.50$ μm,

$n_1 = 1.40$, and $n_2 = 1.60$. (a) In wavelengths, what is their phase difference after they both have emerged from the layers? (b) If the waves later arrive at some common point, what type of interference do they undergo?

10. Two Light Waves In Fig. 36-3, assume that the two light waves, of wavelength 620 nm in air, are initially out of phase by π rad. The indexes of refraction of the media are $n_1 = 1.45$ and $n_2 = 1.65$. (a) What is the least thickness L that will put the waves exactly in phase once they pass through the two media? (b) What is the next greater L that will do this?

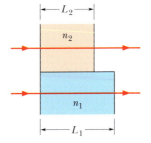

FIGURE 36-20 ▪
Problem 9.

SEC. 36-4 ▪ YOUNG'S INTERFERENCE EXPERIMENT

11. Green Light Monochromatic green light, of wavelength 550 nm, illuminates two parallel narrow slits 7.70 μm apart. Calculate the angular deviation (θ in Fig. 36-8) of the third-order (for $m = 3$) bright fringe (a) in radians and (b) in degrees.

12. Phase Difference What is the phase difference of the waves from the two slits when they arrive at the mth dark fringe in a Young's double-slit experiment?

13. Blue-Green Light Suppose that Young's experiment is performed with blue-green light of wavelength 500 nm. The slits are 1.20 mm apart, and the viewing screen is 5.40 m from the slits. How far apart are the bright fringes.

14. Angular Separation In a double-slit arrangement the slits are separated by a distance equal to 100 times the wavelength of the light passing through the slits. (a) What is the angular separation in radians between the central maximum and an adjacent maximum? (b) What is the distance between these maxima on a screen 50.0 cm from the slits?

15. Interference Fringes A double-slit arrangement produces interference fringes for sodium light ($\lambda = 589$ nm) that have an angular separation of 3.50×10^{-3} rad. For what wavelength would the angular separation be 10.0% greater?

16. Immersed in Water A double-slit arrangement produces interference fringes for sodium light ($\lambda = 589$ nm) that are 0.20° apart. What is the angular fringe separation if the entire arrangement is immersed in water ($n = 1.33$)?

17. Radio Frequency Sources Two radio-frequency point sources separated by 2.0 m are radiating in phase with $\lambda = 0.50$ m. A detector moves in a circular path around the two sources in a plane containing them. Without written calculation, find how many maxima it detects.

18. Long-Range Radio Waves Sources A and B emit long-range radio waves of wavelength 400 m, with the phase of the emission from A ahead of that from source B by 90°. The distance r_A from A to a detector is greater than the corresponding distance r_B by 100 m. What is the phase difference at the detector?

19. Two Interference Patterns In a double-slit experiment the distance between slits is 5.0 nm and the slits are 1.0 m from the screen. Two interference patterns can be seen on the screen: one due to light with wavelength 480 nm, and the other due to light with wavelength 600 nm. What is the separation on the screen between the third-order ($m = 3$) bright fringes of the two interference patterns?

20. Identical Radiators In Fig. 36-21, S_1 and S_2 are identical radiators of waves that are in phase and of the same wavelength λ. The radiators are separated by distance $d = 3.00\lambda$. Find the greatest distance from S_1, along the x axis, for which fully destructive interference occurs. Express this distance in wavelengths.

FIGURE 36-21 ▪
Problems 20 and 27.

21. Mica Flake A thin flake of mica ($n = 1.58$) is used to cover one slit of a double-slit interference arrangement. The central point on the viewing screen is now occupied by what had been the seventh bright side fringe ($m = 7$) before the mica was used. If $\lambda = 550$ nm, what is the thickness of the mica? (*Hint:* Consider the wavelength of the light within the mica.)

22. Laser Light Laser light of wavelength 632.8 nm passes through a double-slit arrangement at the front of a lecture room, reflects off a mirror 20.0 m away at the back of the room, and then produces an interference pattern on a screen at the front of the room. The distance between adjacent bright fringes is 10.0 cm. (a) What is the slit separation? (b) What happens to the pattern when the lecturer places a thin cellophane sheet over one slit, thereby increasing by 2.50 the number of wavelengths along the path that includes the cellophane?

SEC. 36-6 ▪ INTENSITY IN DOUBLE-SLIT INTERFERENCE

23. Same Frequency Two waves of the same frequency have amplitudes 1.00 and 2.00. They interfere at a point where their phase difference is 60.0°. What is the resultant amplitude?

24. Find Sum Find the sum y of the following quantities:

$$y_1 = 0 \sin \omega t \quad \text{and} \quad y_2 = 8.0 \sin(\omega t + 30°).$$

25. Use Phasors Add the quantities

$$y_1 = 10 \sin \omega t$$
$$y_2 = 15 \sin(\omega t + 30°)$$
$$y_3 = 5.0 \sin(\omega t - 45°)$$

using the phasor method.

26. Sketch Intensity Light of wavelength 600 nm is incident normally on two parallel narrow slits separated by 0.60 mm. Sketch the intensity pattern observed on a distant screen as a function of angle θ from the pattern's center for the range of values $0 \le \theta \le 0.0040$ rad.

27. Electromagnetic Waves S_1 and S_2 in Fig. 36-21 are point sources of electromagnetic waves of wavelength 1.00 m. They are in phase and separated by $d = 4.00$ m, and they emit at the same power. (a) If a detector is moved to the right along the x axis from source S_1, at what distances from S_1 are the first three interference maxima detected? (b) Is the intensity of the nearest minimum exactly zero? (*Hint:* Does the intensity of a wave from a point source remain constant with an increase in distance from the source?).

28. Horizontal Arrow The double horizontal arrow in Fig. 36-9 marks the points on the intensity curve where the intensity of the central fringe is half the maximum intensity. Show that the angular separation $\Delta\theta$ between the corresponding points on the viewing screen is

$$\Delta\theta = \frac{\lambda}{2d}$$

if θ in Fig. 36-8 is small enough so that $\sin\theta \approx \theta$.

29. Wider Slit Suppose that one of the slits of a double-slit interference experiment is wider than the other, so the amplitude of the light reaching the central part of the screen from one slit, acting alone, is twice that from the other slit, acting alone. Derive an expression for the light intensity I at the screen as a function of θ, corresponding to Eqs. 36-21 and 36-22.

SEC. 36-7 ■ INTERFERENCE FROM THIN FILMS

30. Reflections In Fig. 36-22, light wave W_1 reflects once from a reflecting surface while light wave W_2 reflects twice from that surface and once from a reflecting sliver at distance L from the mirror. The waves are initially in phase and have a wavelength of 620 nm. Neglect the slight tilt of the rays. (a) For what

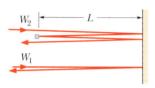

FIGURE 36-22 ■
Problems 30 and 32.

least value of L are the reflected waves exactly out of phase? (b) How far must the sliver be moved to put the waves exactly out of phase again?

31. Bright Light Bright light of wavelength 585 nm is incident perpendicularly on a soap film ($n = 1.33$) of thickness 1.21 μm, suspended in air. Is the light reflected by the two surfaces of the film closer to interfering fully destructively or fully constructively?

32. Exactly Out of Phase Suppose the light waves of Problem 30 are initially exactly out of phase. Find an expression for the values of L (in terms of the wavelength λ) that put the reflected waves exactly in phase.

33. Soap Film Light of wavelength 624 nm is incident perpendicularly on a soap film (with $n = 1.33$) suspended in air. What are the least two thicknesses of the film for which the reflections from the film undergo fully constructive interference?

34. Camera Lens A camera lens with index of refraction greater than 1.30 is coated with a thin transparent film of index of refraction 1.25 to eliminate by interference the reflection of light at wavelength λ that is incident perpendicularly on the lens. In terms of λ, what minimum film thickness is needed?

35. Rhinestones The rhinestones in costume jewelry are glass with index of refraction 1.50. To make them more reflective, they are often coated with a layer of silicon monoxide of index of refraction 2.00. What is the minimum coating thickness needed to ensure that light of wavelength 560 nm and of perpendicular incidence will be reflected from the two surfaces of the coating with fully constructive interference?

36. Five Sections In Fig. 36-23, light of wavelength 600 nm is incident perpendicularly on five sections of a transparent structure suspended in air. The structure has index of refraction 1.50. The thickness of each section is given in terms of $L = 4.00$ μm. For

which sections will the light that is reflected from the top and bottom surfaces of that section undergo fully constructive interference?

37. Coat Glass We wish to coat flat glass ($n = 1.50$) with a transparent material ($n = 1.25$) so that reflection of light at wavelength 600 nm is eliminated by interference. What minimum thickness can the coating have to do this?

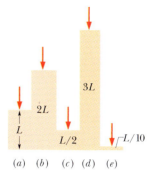

FIGURE 36-23 ■
Problem 36.

38. Four Thin Layers In Fig. 36-24, light is incident perpendicularly on four thin layers of thickness L. The indexes of refraction of the thin layers and of the media above and below these layers are given. Let λ represent the wavelength of the light in air and n_2 represent the index of refraction of the thin layer in each situation. Consider only the transmission of light that undergoes no reflection or two reflections, as in Fig. 36-24a. For which of the situations does the expression

$$\lambda = \frac{2Ln_2}{m}, \qquad \text{for } m = 1, 2, 3, \dots,$$

give the wavelengths of the transmitted light that undergoes fully constructive interference?

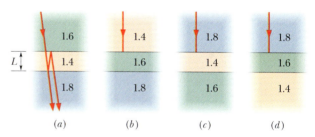

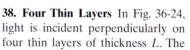

FIGURE 36-24 ■ Problems 38 and 39.

39. Leaking Tanker A disabled tanker leaks kerosene ($n = 1.20$) into the Persian Gulf creating a large slick on top of the water ($n = 1.30$). (a) If you are looking straight down from an airplane, while the Sun is overhead, at a region of the slick where its thickness is 460 nm, for which wavelength(s) of visible light is the reflection brightest because of constructive interference? (b) If you are scuba diving directly under this same region of the slick, for which wavelength(s) of visible light is the transmitted intensity strongest? (*Hint:* Use Fig. 36-24a with appropriate indexes of refraction.)

40. Plane Wave A plane wave of monochromatic light is incident normally on a uniform thin film of oil that covers a glass plate. The wavelength of the source can be varied continuously. Fully destructive interference of the reflected light is observed for wavelengths of 500 and 700 nm and for no wavelengths in between. If the index of refraction of the oil is 1.30 and that of the glass is 1.50, find the thickness of the oil film.

41. Monochromatic Light A plane monochromatic light wave in air is perpendicularly incident on a thin film of oil that covers a glass plate. The wavelength of the source may be varied continuously. Fully destructive interference of the reflected light is observed for wavelengths of 500 and 700 nm and for no wavelength in between. The index of refraction of the glass is 1.50. Show that the index of refraction of the oil must be less than 1.50.

42. Soap Film Two The reflection of perpendicularly incident white light by a soap film in air has an interference maximum at 600 nm and a minimum at 450 nm, with no minimum in between. If $n = 1.33$ for the film, what is the film thickness, assumed uniform?

43. Glass Plates In Fig. 36-25, a broad beam of light of wavelength 683 nm is sent directly downward through the top plate of a pair of glass plates. The plates are 120 mm long, touch at the left end, and are separated by a wire of diameter 0.048 mm at the right end. The air between the plates acts as a thin film. How many bright fringes will be seen by an observer looking down through the top plate?

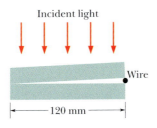

FIGURE 36-25 ▪ Problems 43 and 44.

44. Directly Downward In Fig. 36-25, white light is sent directly downward through the top plate of a pair of glass plates. The plates touch at the left end and are separated by a wire of diameter 0.048 mm at the right end; the air between the plates acts as a thin film. An observer looking down through the top plate sees bright and dark fringes due to that film. (a) Is a dark fringe or a bright fringe seen at the left end? (b) To the right of that end, fully destructive interference occurs at different locations for different wavelengths of the light. Does it occur first for the red end or the blue end of the visible spectrum?

45. Wedge-Shaped A broad beam of light of wavelength 630 nm is incident at 90° on a thin, wedge-shaped film with index of refraction 1.50. An observer intercepting the light transmitted by the film sees 10 bright and 9 dark fringes along the length of the film. By how much does the film thickness change over this length?

46. Acetone A thin film of acetone ($n = 1.25$) coats a thick glass plate ($n = 1.50$). White light is incident normal to the film. In the reflections, fully destructive interference occurs at 600 nm and fully constructive interference at 700 nm. Calculate the thickness of the acetone film.

47. Two Glass Plates Two glass plates are held together at one end to form a wedge of air that acts as a thin film. A broad beam of light of wavelength 480 nm is directed through the plates, perpendicular to the first plate. An observer intercepting light reflected from the plates sees on the plates an interference pattern that is due to the wedge of air. How much thicker is the wedge at the sixteenth bright fringe than it is at the sixth bright fringe, counting from where the plates touch?

48. Broad Beam A broad beam of monochromatic light is directed perpendicularly through two glass plates that are held together at one end to create a wedge of air between them. An observer intercepting light reflected from the wedge of air, which acts as a thin film, sees 4001 dark fringes along the length of the wedge. When the air between the plates is evacuated, only 4000 dark fringes are seen. Calculate the index of refraction of air from these data.

49. Radius of Curvature Figure 36-26a shows a lens with radius of curvature R lying on plane glass plate and illuminated from above by light with wavelength λ. Figure 36-26b (a photograph taken from above the lens) shows that circular interference fringes (called

Newton's rings) appear, associated with the variable thickness d of the air film between the lens and the plate. Find the radii r of the interference maxima assuming $r/R \ll 1$.

50. Newtons's Rings One In a Newton's rings experiment (see Problem 49), the radius of curvature R of the lens is 5.0 m and the lens diameter is 20 mm. (a) How many bright rings are produced? Assume that $\lambda = 589$ nm. (b) How many bright rings would be produced if the arrangement were immersed in water ($n = 1.33$)?

51. Newton's Rings Two A Newton's rings apparatus is to be used to determine the radius of curvature of a lens (see Fig. 36-26 and Problem 49). The radii of the nth and $(n + 20)$th bright rings are measured and found to be 0.162 and 0.368 cm, respectively, in light of wavelength 546 nm. Calculate the radius of curvature of the lower surface of the lens.

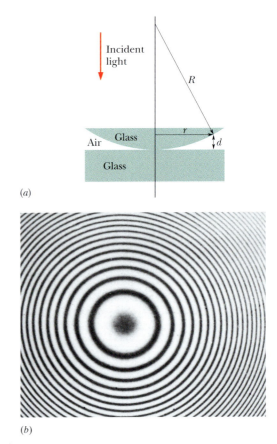

(a)

(b)

FIGURE 36-26 ▪ Problems 49 through 52.

52. Newton's Rings Three (a) Use the result of Problem 49 to show that, in a Newton's rings experiment, the difference in radius between adjacent bright rings (maxima) is given by

$$\Delta r = r_{m+1} - r_m \approx \tfrac{1}{2}\sqrt{\lambda R/m},$$

assuming $m \gg 1$. (b) Now show that the *area* between adjacent bright rings is given by

$$A = \pi\lambda R,$$

assuming $m \gg 1$. Note that this area is independent of m.

53. Microwave Transmitter In Fig. 36-27, a microwave transmitter at height a above the water level of a wide lake transmits microwaves of wavelength λ toward a receiver on the opposite shore, a distance x above the water level. The microwaves reflecting from the water interfere with the microwaves arriving directly from the transmitter. Assuming that the lake width D is much greater than a and x, and that $\lambda \geq a$, at what values of x is the signal at the receiver maximum? (*Hint:* Does the reflection cause a phase change?)

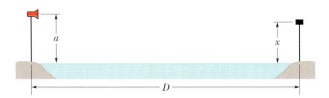

FIGURE 36-27 ▪ Problem 53.

SEC. 36-8 ▪ MICHELSON'S INTERFEROMETER

54. Thin Film A thin film with index of refraction $n = 1.40$ is placed in one arm of a Michelson interferometer, perpendicular to the optical path. If this causes a shift of 7.0 fringes of the pattern produced by light of wavelength 589 nm, what is the film thickness?

55. Move the Mirror If mirror M_2 in a Michelson interferometer (Fig. 36-17) is moved through 0.233 mm, a shift of 792 fringes occurs. What is the wavelength of the light producing the fringe pattern?

56. Light at Two Wavelengths The element sodium can emit light at two wavelengths, $\lambda_1 = 589.10$ nm and $\lambda_2 = 589.59$ nm. Light from sodium is being used in a Michelson interferometer (Fig. 36-17).

Through what distance must mirror M_2 be moved to shift the fringe pattern for one wavelength by 1.00 fringe more than the fringe pattern for the other wavelength?

57. Airtight Chamber In Fig. 36-28, an airtight chamber 5.0 cm long with glass windows is placed in one arm of a Michelson interferometer. Light of wavelength $\lambda = 500$ nm is used. Evacuating the air from the chamber causes a shift of 60 fringes. From these data, find the index of refraction of air at atmospheric pressure.

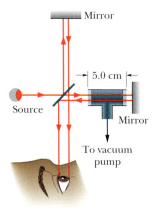

FIGURE 36-28 ▪ Problem 57.

58. Observed Intensity Write an expression for the intensity observed in a Michelson interferometer (Fig. 36-17) as a function of the position of the moveable mirror. Measure the position of the mirror from the point at which $d_2 = d_1$.

Additional Problems

59. Arranging the Patio Speakers You have set up two stereo speakers on your patio as shown in the top view diagram in Fig. 36-29. You are worried that at certain positions you will lose frequencies as a result of interference. The coordinate grid on the edge of the picture has its large tick marks separated by 1 m. For ease of calculation, make the following assumptions:

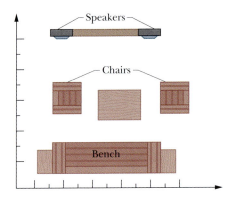

FIGURE 36-29 ■ Problem 59.

- Assume that the relevant objects lie on integer or half-integer grid points of the coordinate system.
- Take the speed of sound to be 343 m/s.
- Ignore the reflection of sound from the house, trees, etc.
- Assume that the speakers are in phase.

(a) What will happen if you are sitting in the middle of the bench?

(b) If you are sitting in the lawn chair on the left, what will be the lowest frequency you will lose to destructive interference?

(c) Can you restore the frequency lost in part (b) by switching the leads to one of the speakers, thereby reversing the phase of that source?

(d) With the leads reversed, what will happen to the sound for a person sitting at the center of the bench?

60. What Happens If a Double Slit Winks? When a laser beam is incident on a double slit, a closeup of the center of the pattern looks like that shown in Fig. 36-30. If one of the slits is covered (the left one) but the other slit remains open, what will the pattern look like? Explain how you know.

FIGURE 36-30 ■ Problem 60.

37 | Diffraction

Georges Seurat painted *Sunday Afternoon on the Island of La Grande Jatte* using not brush strokes in the usual sense, but rather a myriad of small colored dots, in a style of painting now known as pointillism. You can see the dots if you stand close enough to the painting, but as you move away from it, they eventually blend and cannot be distinguished. Moreover, the color that you see at any given place on the painting changes as you move away—which is why Seurat painted with the dots.

What causes this change in color?

The answer is in this chapter.

FIGURE 37-1 ■ This diffraction pattern appeared on a viewing screen when light that had passed through a narrow but tall vertical slit reached the screen. Diffraction causes light to flare out perpendicular to the long sides of the slit. That produces an interference pattern consisting of a broad central maximum less intense and narrower secondary (or side) maxima, with minima between them.

FIGURE 37-2 ■ The diffraction pattern produced by a razor blade in monochromatic light. Note the lines of alternating maximum and minimum intensity.

37-1 Diffraction and the Wave Theory of Light

In Chapter 36 we defined diffraction rather loosely as the flaring of light as it emerges from a narrow slit. More than just flaring occurs, however, because the light produces an interference pattern called a **diffraction pattern.** For example, when monochromatic light from a distant source (or a laser) passes through a narrow slit and is then intercepted by a viewing screen, the light produces on the screen a diffraction pattern like that in Fig. 37-1. This pattern consists of a broad and intense (very bright) central maximum and a number of narrower and less intense maxima (called **secondary** or **side** maxima) to both sides. In between the maxima are minima.

Such a pattern would be totally unexpected in geometrical optics: If light traveled in straight lines as rays, then the slit would allow some of those rays through and they would form a sharp, bright rendition of the slit on the viewing screen. As in Chapter 36, we again must conclude that geometrical optics is only an approximation.

Diffraction of light is not limited to situations of light passing through a narrow opening (such as a slit or pinhole). It also occurs when light passes an edge, such as the edges of the razor blade whose diffraction pattern is shown in Fig. 37-2. Note the lines of maxima and minima that run approximately parallel to the edges, at both the inside edges of the blade and the outside edges. As the light passes, say, the vertical edge at the left, it flares left and right and undergoes interference, producing the pattern along the left edge. The rightmost portion of that pattern actually lies within what would have been the shadow of the blade if geometrical optics prevailed.

You encounter a common example of diffraction when you look at a clear blue sky and see tiny specks and hair-like structures floating in your view. These *floaters,* as they are called, are produced when light passes the edges of tiny deposits in the vitreous humor, the transparent material filling most of your eyeball. What you are seeing when a floater is in your field of vision is the diffraction pattern produced on the retina by one of these deposits. If you sight through a pinhole in an otherwise opaque sheet so as to make the light entering your eye approximately a plane wave, you can distinguish individual maxima and minima in the patterns.

The Fresnel Bright Spot

Diffraction finds a ready explanation in the wave theory of light. However, this theory, originally advanced in the late 1600s by Huygens and used 123 years later by Young to explain double-slit interference, was very slow in being adopted, largely because it ran counter to Newton's theory that light was a stream of particles.

Newton's view was the prevailing view in French scientific circles of the early 19th century, when Augustin Fresnel was a young military engineer. Fresnel, who believed in the wave theory of light, submitted a paper to the French Academy of Sciences describing his experiments with light and his wave-theory explanations of them.

In 1819, the Academy, dominated by supporters of Newton and thinking to challenge the wave point of view, organized a prize competition for an essay on the subject of diffraction. Fresnel won. The Newtonians, however, were neither converted nor silenced. One of them, S. D. Poisson, pointed out the "strange result" that if Fresnel's theories were correct, then light waves should flare into the shadow region of a sphere as they pass the edge of the sphere, producing a bright spot at the center of the shadow. The prize committee arranged to have Dominique Argo test the famous mathematician's prediction. He discovered (see Fig. 37-3) that the predicted *Fresnel bright spot,* as we call it today, was indeed there!* Nothing builds confidence in a

---

* Since Poisson predicted the spot and Argo discovered it, an alternate name is the Poisson-Argo bright spot.

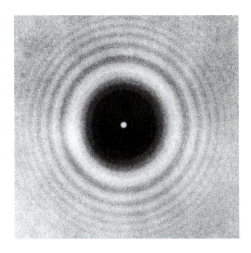

FIGURE 37-3 ◼ A photograph of the diffraction pattern of a disk. Note the concentric diffraction rings and the Fresnel bright spot at the center of the pattern. This experiment is essentially identical to that arranged by the committee testing Fresnel's theories, because both the sphere they used and the disk used here have a cross section with a circular edge.

theory so much as having one of its unexpected and counterintuitive predictions verified by experiment.

37-2 Diffraction by a Single Slit: Locating the Minima

Let us now examine the diffraction pattern of plane waves of light of wavelength λ that are diffracted by a single, long, narrow slit of width a in an otherwise opaque screen B, as shown in cross section in Fig. 37-4a. (In that figure, the slit's length extends into and out of the page, and the incoming wavefronts are parallel to screen B.) When the diffracted light reaches viewing screen C, waves from different points within the slit undergo interference and produce a diffraction pattern of bright and dark fringes (interference maxima and minima) on the screen. To locate the fringes, we shall use a procedure somewhat similar to the one we used to locate the fringes in a two-slit interference pattern. However, diffraction is more mathematically challenging, and here we shall be able to find equations for only the dark fringes.

Before we do that, however, we can justify the central bright fringe seen in Fig. 37-1 by noting that the Huygens wavelets from all points in the slit travel about the same distance to reach the center of the pattern and thus are in phase there. As for the other bright fringes, we can say only that they are approximately halfway between adjacent dark fringes.

To find the dark fringes, we shall use a clever (and simplifying) strategy that involves pairing up all the rays coming through the slit and then finding what conditions cause the wavelets of the rays in each pair to cancel each other. Figure 37-4a shows how we apply this strategy to locate the first dark fringe, at point P_1. First, we mentally divide the slit into two zones of equal widths $a/2$. Then we extend to P_1 a light ray r_1 from the top point of the top zone and a light ray r_2 from the top point of the bottom zone. A central axis is drawn from the center of the slit to screen C, and P_1 is located at an angle θ to that axis.

The wavelets of the pair of rays r_1 and r_2 are in phase within the slit because they originate from the same wavefront passing through the slit, along the width of the slit. However, to produce the first dark fringe they must be out of phase by $\lambda/2$ when they reach P_1; this phase difference is due to their path length difference, with the wavelet of r_2 traveling a longer path to reach P_1 than the wavelet of r_1. To display this path length difference, we find a point b on ray r_2 such that the path length from b to P_1 matches the path length of ray r_1. Then the path length difference between the two rays is the distance from the center of the slit to b.

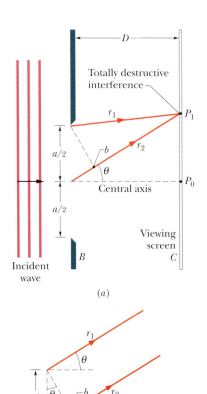

FIGURE 37-4 ◼ (a) Waves from the top points of two zones of width $a/2$ undergo totally destructive interference at point P_1 on viewing screen C. (b) For $D \gg a$, we can approximate rays r_1 and r_2 as being parallel, at angle θ to the central axis.

When viewing screen C is near screen B, as in Fig. 37-4a, the diffraction pattern on C is difficult to describe mathematically. However, we can simplify the mathematics considerably if we arrange for the distance between the slit and screen D to be much larger than the slit width a. Then we can approximate rays r_1 and r_2 as being parallel, at angle θ to the central axis (Fig. 37-4b). We can also approximate the triangle formed by point b, the top point of the slit, and the center point of the slit as being a right triangle, and one of the angles inside that triangle as being θ. The path length difference between rays r_1 and r_2 (which is still the distance from the center of the slit to point b) is then equal to $(a/2) \sin\theta$.

We can repeat this analysis for any other pair of rays originating at corresponding points in the two zones (say, at the midpoints of the zones) and extending to point P_1. Each such pair of rays has the same path length difference $(a/2) \sin\theta$. Setting this common path length difference equal to $\lambda/2$ (our condition for the first dark fringe), we have

$$\frac{a}{2} \sin\theta = \frac{\lambda}{2},$$

which gives us

$$a \sin\theta = \lambda \qquad \text{(first minimum for } D \gg a\text{)}. \qquad (37\text{-}1)$$

Given slit width a and wavelength λ, Eq. 37-1 tells us the angle θ of the first dark fringe above and (by symmetry) below the central axis.

Note that if we begin with $a > \lambda$ and then narrow the slit while holding the wavelength constant, we increase the angle at which the first dark fringes appear; that is, the extent of the diffraction (the extent of the flaring and the width of the pattern) is *greater* for a *narrower* slit. When we have reduced the slit width to the wavelength (that is, $a = \lambda$), the angle of the first dark fringes is $90°$. Since the first dark fringes mark the two edges of the central bright fringe, that bright fringe must then cover the entire viewing screen.

We find the second dark fringes above and below the central axis as we found the first dark fringes, except that we now divide the slit into *four* zones of equal widths $a/4$, as shown in Fig. 37-5a. We then extend rays r_1, r_2, r_3, and r_4 from the top points of the zones to point P_2, the location of the second dark fringe above the central axis. To produce that fringe, the path length difference between r_1 and r_2, that between r_2 and r_3, and that between r_3 and r_4 must all be equal to $\lambda/2$.

For $D \gg a$, we can approximate these four rays as being parallel, at angle θ to the central axis. To display their path length differences, we extend a perpendicular line through each adjacent pair of rays, as shown in Fig. 37-5b, to form a series of right triangles, each of which has a path length difference as one side. We see from the top triangle that the path length difference between r_1 and r_2 is $(a/4) \sin\theta$. Similarly, from the bottom triangle, the path length difference between r_3 and r_4 is also $(a/4) \sin\theta$. In fact, the path length difference for any two rays that originate at corresponding points in two adjacent zones is $(a/4) \sin\theta$. Since in each such case the path length difference is equal to $\lambda/2$, we have

$$\frac{a}{4} \sin\theta = \frac{\lambda}{2},$$

which gives us

$$a \sin\theta = 2\lambda \qquad \text{(second minimum for } D \gg a\text{)}. \qquad (37\text{-}2)$$

We could now continue to locate dark fringes in the diffraction pattern by splitting up the slit into more zones of equal width. We would always choose an even num-

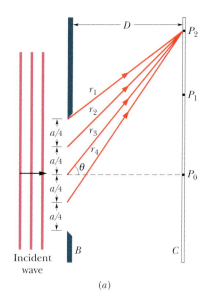

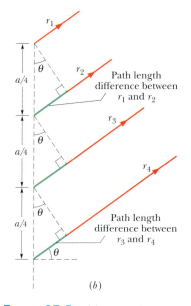

FIGURE 37-5 ■ (a) Waves from the top points of four zones of width $a/4$ undergo totally destructive interference at point P_2. (b) For $D \gg a$, we can approximate rays $r_1, r_2, r_3,$ and r_4 as being parallel, at angle θ to the central axis.

ber of zones so that the zones (and their waves) could be paired as we have been doing. We would find that the dark fringes above and below the central axis can be located with the following general equation:

$$a \sin\theta = m\lambda, \quad \text{for } m = 1, 2, 3, \ldots \quad \text{(single slit minima—dark fringes).} \quad (37\text{-}3)$$

You can remember this result in the following way. Draw a triangle like the one in Fig. 37-4b, but for the full slit width a, and note that the path length difference between the top and bottom rays from the slit equals $a \sin\theta$. Thus, Eq. 37-3 says:

> In a single-slit diffraction experiment, dark fringes are produced where the path length differences ($a \sin\theta$) between the top and bottom rays are equal to $\lambda, 2\lambda, 3\lambda \ldots$.

This may seem to be wrong, because the waves of those two particular rays will be exactly in phase with each other when their path length difference is an integer number of wavelengths. However, they each will still be part of a pair of waves that are exactly out of phase with each other; thus, *each* will be canceled by some other wave.

READING EXERCISE 37-1: We produce a diffraction pattern on a viewing screen by means of a long narrow slit illuminated by blue light. Does the pattern expand away from the bright center (the maxima and minima shift away from the center) or contract toward it if we (a) switch to yellow light or (b) decrease the slit width? ■

TOUCHSTONE EXAMPLE 37-1: White Light, Red Light

A slit of width a is illuminated by white light (which consists of all the wavelengths in the visible range).

(a) For what value of a will the first minimum for red light of wavelength $\lambda = 650$ nm appear at $\theta = 15°$?

SOLUTION ■ The **Key Idea** here is that diffraction occurs separately for each wavelength in the range of wavelengths passing through the slit, with the locations of the minima for each wavelength given by Eq. 37-3 ($a \sin\theta = m\lambda$). When we set $m = 1$ (for the first minimum) and substitute the given values of θ and λ, Eq. 37-3 yields

$$a = \frac{m\lambda}{\sin\theta} = \frac{(1)(650 \text{ nm})}{\sin 15°}$$

$$= 2511 \text{ nm} \approx 2.5 \ \mu\text{m}. \quad \text{(Answer)}$$

For the incident light to flare out that much ($\pm 15°$ to the first minima) the slit has to be very fine indeed—about four times the wavelength. For comparison, note that a fine human hair may be about 100 μm in diameter.

(b) What is the wavelength λ' of the light whose first side diffraction maximum is at 15°, thus coinciding with the first minimum for the red light?

SOLUTION ■ The **Key Idea** here is that the first side maximum for any wavelength is about halfway between the first and second minima for that wavelength. Those first and second minima can be located with Eq. 37-3 by setting $m = 1$ and $m = 2$, respectively. Thus, the first side maximum can be located *approximately* by setting $m = 1.5$. Then Eq. 37-3 becomes

$$a \sin\theta = 1.5\lambda'.$$

Solving for λ' and substituting known data yield

$$\lambda' = \frac{a \sin\theta}{1.5} = \frac{(2511 \text{ nm})(\sin 15°)}{1.5}$$

$$= 430 \text{ nm}. \quad \text{(Answer)}$$

Light of this wavelength is violet. The first side maximum for light of wavelength 430 nm will always coincide with the first minimum for light of wavelength 650 nm, no matter what the slit width is. If the slit is relatively narrow, the angle θ at which this overlap occurs will be relatively large, and conversely for a wide slit the angle is small.

37-3 Intensity in Single-Slit Diffraction, Qualitatively

In Section 37-2 we saw how to find the positions of the minima and the maxima in a single-slit diffraction pattern. Now we turn to a more general problem: Find an expression for the intensity I of the pattern as a function of θ, the angular position of a point on a viewing screen.

To do this, we divide the slit of Fig. 37-4a into N zones of equal widths Δx small enough that we can assume each zone acts as a source of Huygens wavelets. We wish to superimpose the wavelets arriving at an arbitrary point P on the viewing screen, at angle θ to the central axis, so that we can determine the amplitude E_θ of the magnitude of the electric field of the resultant wave at P. The intensity of the light at P is then proportional to the square of that amplitude.

To find E_θ, we need the phase relationships among the arriving wavelets. The phase difference between wavelets from adjacent zones is given by

$$(\text{phase difference}) = \left(\frac{2\pi}{\lambda}\right)(\text{path length difference}).$$

For point P at angle θ, the path length difference between wavelets from adjacent zones is $\Delta x \sin\theta$, so the phase difference $\Delta\phi$ between wavelets from adjacent zones is

$$\Delta\phi = \left(\frac{2\pi}{\lambda}\right)(\Delta x \sin\theta). \tag{37-4}$$

We assume that the wavelets arriving at P all have the same amplitude ΔE. To find the amplitude E_θ of the resultant wave at P, we add the amplitudes ΔE via phasors. To do this, we construct a diagram of N phasors, one corresponding to the wavelet from each zone in the slit.

For point P_0 at $\theta = 0$ on the central axis of Fig. 37-4a, Eq. 37-4 tells us that the phase difference $\Delta\phi$ between the wavelets is zero; that is, the wavelets all arrive in phase. Figure 37-6a is the corresponding phasor diagram; adjacent phasors represent wavelets from adjacent zones and are arranged head to tail. Because there is zero phase difference between the wavelets, there is zero angle between each pair of adjacent phasors. The amplitude E_θ of the net wave at P_θ is the vector-like sum of these phasors. This arrangement of the phasors turns out to be the one that gives the greatest value for the amplitude E_θ. We call this value E^{max}; that is, E^{max} is the value of E_θ for $\theta = 0$.

We next consider a point P that is at a small angle θ to the central axis. Equation 37-4 now tells us that the phase difference $\Delta\phi$ between wavelets from adjacent zones is no longer zero. Figure 37-6b shows the corresponding phasor diagram; as before,

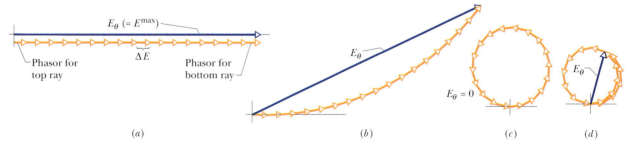

$E_\theta (= E^{\text{max}})$

ΔE

Phasor for top ray

Phasor for bottom ray

E_θ

$E_\theta = 0$

E_θ

(a)   (b)   (c)   (d)

FIGURE 37-6 ■ Phasor diagrams for $N = 18$ phasors, corresponding to the division of a single slit into 18 zones. Resultant amplitudes E_θ are shown for (a) the central maximum at $\theta = 0$, (b) a point on the screen lying at a small angle θ to the central axis, (c) the first minimum, and (d) the first side maximum.

the phasors are arranged head to tail, but now there is an angle $\Delta\phi$ between adjacent phasors. The amplitude E_θ at this new point is still the vector sum of the phasors, but it is smaller than the amplitude in Fig. 37-6a, which means that the intensity of the light is less at this new point P than at P_θ.

If we continue to increase θ, the angle $\Delta\phi$ between adjacent phasors increases, and eventually the chain of phasors curls completely around so that the head of the last phasor just reaches the tail of the first phasor (Fig. 37-6c). The amplitude E_θ is now zero, which means that the intensity of the light is also zero. We have reached the first minimum, or dark fringe, in the diffraction pattern. The first and last phasors now have a phase difference of 2π rad, which means that the path length difference between the top and bottom rays through the slit equals one wavelength. Recall that this is the condition we determined for the first diffraction minimum.

As we continue to increase θ, the angle $\Delta\phi$ between adjacent phasors continues to increase, the chain of phasors begins to wrap back on itself, and the resulting coil begins to shrink. Amplitude E_θ now increases until it reaches a maximum value in the arrangement shown in Fig. 37-6d. This arrangement corresponds to the first side maximum in the diffraction pattern.

If we increase θ a bit more, the resulting shrinkage of the coil decreases E_θ, which means that the intensity also decreases. When θ is increased enough, the head of the last phasor again meets the tail of the first phasor. We have then reached the second minimum.

We could continue this qualitative method of determining the maxima and minima of the diffraction pattern but, instead, we shall now turn to a quantitative method.

READING EXERCISE 37-2: The figures represent, in smoother form (with more phasors) than Fig. 37-6, the phasor diagrams for two points of a diffraction pattern that are on opposite sides of a certain diffraction maximum. (a) Which maximum is it? (b) What is the approximate value of m (in Eq. 37-3) that corresponds to this maximum?

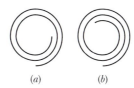

(a)   (b)

37-4 Intensity in Single-Slit Diffraction, Quantitatively

Equation 37-3 tells us how to locate the minima of the single-slit diffraction pattern on screen C of Fig. 37-4a as a function of the angle θ in that figure. Here we wish to derive an expression for the intensity I_θ of the pattern as a function of θ. We state, and shall prove below, that the intensity is given by

$$I_\theta = I^{max}\left(\frac{\sin\alpha}{\alpha}\right)^2,\qquad(37\text{-}5)$$

where

$$\alpha = \frac{1}{2}\Delta\phi = \frac{\pi a}{\lambda}\sin\theta.\qquad(37\text{-}6)$$

The symbol α is just a convenient connection between the angle θ that locates a point on the viewing screen and the light intensity I_θ at that point. I^{max} is the greatest value of the intensity I_θ in the pattern and occurs at the central maximum (where $\theta = 0$), and $\Delta\phi$ is the phase difference (in radians) between the top and bottom rays from the slit width a.

Study of Eq. 37-5 shows that intensity minima will occur where

$$\alpha = m\pi, \qquad \text{for } m = 1, 2, 3, \ldots.\qquad(37\text{-}7)$$

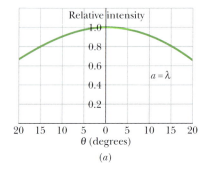

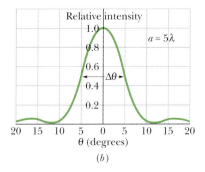

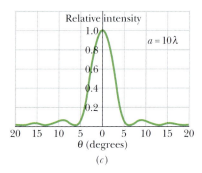

FIGURE 37-7 ■ The relative intensity in single-slit diffraction for three values of the ratio a/λ. The wider the slit is, the narrower is the central diffraction maximum.

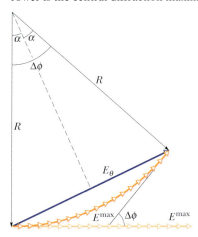

FIGURE 37-8 ■ A construction used to calculate the intensity in single-slit diffraction. The situation shown corresponds to that of Fig. 37-6b.

If we put this result into Eq. 37-6 we find

$$m\pi = \frac{\pi a}{\lambda}\sin\theta, \qquad \text{for } m = 1, 2, 3, \ldots,$$

or
$$a\sin\theta = m\lambda, \qquad \text{for } m = 1, 2, 3, \ldots \quad \text{(minima—dark fringes)}, \qquad (37\text{-}8)$$

which is exactly Eq. 37-3, the expression that we derived earlier for the location of the minima.

Figure 37-7 shows plots of the intensity of a single-slit diffraction pattern, calculated with Eqs. 37-5 and 37-6 for three slit widths: $a = \lambda$, $a = 5\lambda$, and $a = 10\lambda$. Note that as the slit width increases (relative to the wavelength), the width of the *central diffraction maximum* (the central hill-like region of the graphs) decreases; that is, the light undergoes less flaring by the slit. The secondary maxima also decrease in width (and become weaker). In the limit of slit width a being much greater than wavelength λ, the secondary maxima due to the slit disappear; we then no longer have single-slit diffraction (but we still have diffraction due to the edges of the wide slit, like that produced by the edges of the razor blade in Fig. 37-2).

Proof of Eqs. 37-5 and 37-6

The arc of phasors in Fig. 37-8 represents the wavelets that reach an arbitrary point P on the viewing screen of Fig. 37-4, corresponding to a particular small angle θ. The amplitude E_θ of the resultant wave at P is the vector sum of these phasors. If we divide the slit of Fig. 37-4 into infinitesimal zones of width Δx, the arc of phasors in Fig. 37-8 approaches the arc of a circle; we call its radius R as indicated in that figure. The length of the arc must be E^{max}, the amplitude at the center of the diffraction pattern, because if we straightened out the arc we would have the phasor arrangement of Fig. 37-6a (shown lightly in Fig. 37-8).

The angle $\Delta\phi$ in the lower part of Fig. 37-8 is the difference in phase between the infinitesimal vectors at the left and right ends of arc E^{max}. From the geometry, $\Delta\phi$ is also the angle between the two radii marked R in Fig. 37-8. The dashed line in that figure, which bisects $\Delta\phi$, then forms two congruent right triangles. From either triangle we can write

$$\sin\tfrac{1}{2}\Delta\phi = \frac{E_\theta}{2R}. \qquad (37\text{-}9)$$

In radian measure, $\Delta\phi$ is (with E^{max} considered to be a circular arc)

$$\Delta\phi = \frac{E^{\text{max}}}{R}.$$

Solving this equation for R, substituting the result into Eq. 37-9 and re-arranging terms yields

$$E_\theta = \frac{E^{\text{max}}}{\tfrac{1}{2}\Delta\phi}\sin\tfrac{1}{2}\Delta\phi. \qquad (37\text{-}10)$$

In Section 34-4 we saw that the intensity of an electromagnetic wave is proportional to the square of the amplitude of its electric field. Here, this means that the maximum intensity I^{max} (which occurs at the center of the diffraction pattern) is proportional to $(E^{\text{max}})^2$ and the intensity I_θ at angle θ is proportional to E_θ^2. Thus, we may write

$$\frac{I_\theta}{I^{\text{max}}} = \frac{E_\theta^2}{(E^{\text{max}})^2}. \tag{37-11}$$

Substituting for E_θ with Eq. 37-10 and then substituting $\alpha = \frac{1}{2}\Delta\phi$, we are led to the following expression for the intensity as a function of θ:

$$I_\theta = I^{\text{max}}\left(\frac{\sin\alpha}{\alpha}\right)^2.$$

This is exactly Eq. 37-5, one of the two equations we set out to prove.

The second equation we wish to prove relates α to θ: The phase difference $\Delta\phi$ between the rays from the top and bottom of the entire slit may be related to a path length difference with Eq. 37-4; it tells us that

$$\Delta\phi = \left(\frac{2\pi}{\lambda}\right)(a\sin\theta),$$

where a is the sum of the widths Δx of the infinitesimal zones. However, $\Delta\phi = 2\alpha$, so this equation reduces to Eq. 37-6.

READING EXERCISE 37-3:
Two wavelengths, 650 and 430 nm, are used separately in a single-slit diffraction experiment. The figure shows the results as graphs of intensity I versus angle θ for the two diffraction patterns. If both wavelengths are then used simultaneously, what color will be seen in the combined diffraction pattern at (a) angle A and (b) angle B?

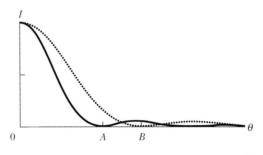

TOUCHSTONE EXAMPLE 37-2: Maxima Intensities

Find the intensities of the first three secondary maxima (side maxima) in the single-slit diffraction pattern of Fig. 37-1, measured relative to the intensity of the central maximum.

SOLUTION ■ One **Key Idea** here is that the secondary maxima lie approximately halfway between the minima, whose angular locations are given by Eq. 37-7 ($\alpha = m\pi$). The locations of the secondary maxima are then given (approximately) by

$$\alpha = (m + \tfrac{1}{2})\pi, \qquad \text{for } m = 1, 2, 3, \ldots,$$

with α in radian measure.

A second **Key Idea** is that we can relate the intensity I at any point in the diffraction pattern to the intensity I^{max} of the central maximum via Eq. 37-5. Thus, we can substitute the approximate values of α for the secondary maxima into Eq. 37-5 to obtain the relative intensities at those maxima. We get

$$\frac{I}{I^{\text{max}}} = \left(\frac{\sin\alpha}{\alpha}\right)^2 = \left(\frac{\sin(m + \tfrac{1}{2})\pi}{(m + \tfrac{1}{2})\pi}\right)^2, \qquad \text{for } m = 1, 2, 3, \ldots.$$

The first of the secondary maxima occurs for $m = 1$, and its relative intensity is

$$\frac{I_1}{I^{\text{max}}} = \left(\frac{\sin(1 + \tfrac{1}{2})\pi}{(1 + \tfrac{1}{2})\pi}\right)^2 = \left(\frac{\sin 1.5\pi}{1.5\pi}\right)^2$$
$$= 4.50 \times 10^{-2} \approx 4.5\%. \tag{Answer}$$

For $m = 2$ and $m = 3$ we find that

$$\frac{I_2}{I^{\text{max}}} = 1.6\% \qquad \text{and} \qquad \frac{I_3}{I^{\text{max}}} = 0.83\%. \tag{Answer}$$

Successive secondary maxima decrease rapidly in intensity. Figure 37-1 was deliberately overexposed to reveal them.

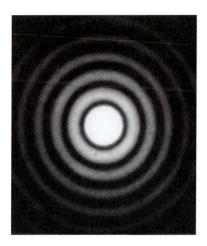

FIGURE 37-9 ▪ The diffraction pattern of a circular aperture. Note the central maximum and the circular secondary maxima. The figure has been overexposed to bring out these secondary maxima, which are much less intense than the central maximum.

37-5 Diffraction by a Circular Aperture

Here we consider diffraction by a circular aperture—that is, a circular opening such as a circular lens, through which light can pass. Figure 37-9 shows the image of a distant point source of light (a star, for instance) formed on photographic film placed in the focal plane of a converging lens. This image is not a point, as geometrical optics would suggest, but a circular disk surrounded by several progressively fainter secondary rings. Comparison with Fig. 37-1 leaves little doubt that we are dealing with a diffraction phenomenon. Here, however, the aperture is a circle of diameter d rather than a rectangular slit.

The analysis of such patterns is complex. It shows, however, that the first minimum for the diffraction pattern of a circular aperture of diameter d is located by

$$\sin\theta = 1.22 \frac{\lambda}{d} \qquad \text{(first minimum—circular aperture).} \qquad (37\text{-}12)$$

The angle θ here is the angle from the central axis to any point on that (circular) minimum. Compare this with Eq. 37-1,

$$\sin\theta = \frac{\lambda}{a} \qquad \text{(first minimum—single slit),} \qquad (37\text{-}13)$$

which locates the first minimum for a long narrow slit of width a. The main difference is the factor 1.22, which enters because of the circular shape of the aperture.

Resolvability

The fact that lens images are diffraction patterns is important when we wish to *resolve* (distinguish) two distant point objects whose angular separation is small. Figure 37-10 shows, in three different cases, the visual appearance and corresponding intensity pattern for two distant point objects (stars, say) with small angular separation. In Figure 37-10a, the objects are not resolved because of diffraction; that is, their diffraction patterns (mainly their central maxima) overlap so much that the two objects cannot be distinguished from a single point object. In Fig. 37-10b the objects are barely resolved, and in Fig. 37-10c they are fully resolved.

FIGURE 37-10 ▪ At the top, the images of two point sources (stars), formed by a converging lens. At the bottom, representations of the image intensities. In (a) the angular separation of the sources is too small for them to be distinguished; in (b) they can be marginally distinguished, and in (c) they are clearly distinguished. Rayleigh's criterion is just satisfied in (b), with the central maximum of one diffraction pattern coinciding with the first minimum of the other.

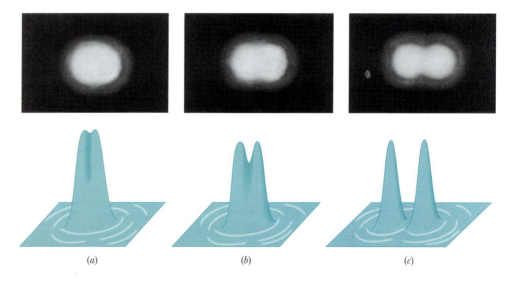

(a)   (b)   (c)

In Fig. 37-10b the angular separation of the two point sources is such that the central maximum of the diffraction pattern of one source is centered on the first minimum of the diffraction pattern of the other, a condition called **Rayleigh's criterion** for resolvability. From Eq. 37-12, two objects that are barely resolvable by this criterion must have an angular separation θ_R of

$$\theta_R = \sin^{-1}\frac{1.22\lambda}{d}.$$

Since the angles involved are small, we can replace $\sin \theta_R$ with θ_R expressed in radians:

$$\theta_R = 1.22\frac{\lambda}{d} \qquad \text{(Rayleigh's criterion—circular aperture)}. \qquad (37\text{-}14)$$

Rayleigh's criterion for resolvability is only an approximation, because resolvability depends on many factors, such as the relative brightness of the sources and their surroundings, turbulence in the air between the sources and the observer, and the functioning of the observer's visual system. Experimental results show that the least angular separation that can actually be resolved by a person is generally somewhat greater than the value given by Eq. 37-14. However, for the sake of calculations here, we shall take Eq. 37-14 as being a precise criterion: If the angular separation θ between the sources is greater than θ_R, we can resolve the sources; if it is less, we cannot.

Rayleigh's criterion can explain the colors in Seurat's *Sunday Afternoon on the Island of La Grande Jatte* (or any other pointillistic painting). When you stand close enough to the painting, the angular separations θ of adjacent dots are greater than θ_R and thus the dots can be seen individually. Their colors are the colors of the paints Seurat used. However, when you stand far enough from the painting, the angular separations θ are less than θ_R and the dots cannot be seen individually. The resulting blend of colors coming into your eye from any group of dots can then cause your brain to "make up" a color for that group—a color that may not actually exist in the group. In this way, Seurat uses your visual system to create the colors of his art.

When we wish to use a lens instead of our visual system to resolve objects of small angular separation, it is desirable to make the diffraction pattern as small as possible. According to Eq. 37-14, this can be done either by increasing the lens diameter or by using light of a shorter wavelength.

For this reason ultraviolet light is often used with microscopes; because of its shorter wavelength, it permits finer detail to be examined than would be possible for the same microscope operated with visible light. It turns out that under certain circumstances, a beam of electrons behaves like a wave. In an *electron microscope* such beams may have an effective wavelength that is 10^{-5} of the wavelength of visible light. They permit the detailed examination of tiny structures, like that in Fig. 37-11, that would be blurred by diffraction if viewed with an optical microscope.

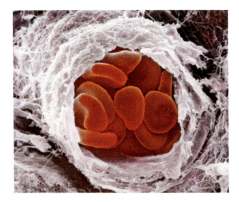

FIGURE 37-11 ■ A false-color scanning electron micrograph of red blood cells traveling through an arterial branch.

READING EXERCISE 37-4: Suppose you can barely resolve two red dots, due to diffraction by the pupil of your eye. If we increase the general illumination around you so that the pupil decreases in diameter, does the resolvability of the dots improve or diminish? Consider only diffraction. (You might experiment to check your answer.) ■

TOUCHSTONE EXAMPLE 37-3: Circular Converging Lens

A circular converging lens, with diameter $d = 32$ mm and focal length $f = 24$ cm, forms images of distant point objects in the focal plane of the lens. Light of wavelength $\lambda = 550$ nm is used.

(a) Considering diffraction by the lens, what angular separation must two distant point objects have to satisfy Rayleigh's criterion?

SOLUTION ■ Figure 37-12 shows two distant point objects P_1 and P_2, the lens, and a viewing screen in the focal plane of the lens. It also shows, on the right, plots of light intensity I versus position on the screen for the central maxima of the images formed by the lens. Note that the angular separation θ_o of the objects equals the angular separation θ_i of the images. Thus, the **Key Idea** here is that if the images are to satisfy Rayleigh's criterion for resolvability, the angular separations on both sides of the lens must be given by Eq. 37-14 (assuming small angles). Substituting the given data, we obtain from Eq. 37-14

$$\theta_o = \theta_i = \theta_R = 1.22 \frac{\lambda}{d}$$
$$= \frac{(1.22)(550 \times 10^{-9}\,\text{m})}{32 \times 10^{-3}\,\text{m}} = 2.1 \times 10^{-5}\,\text{rad.} \quad \text{(Answer)}$$

At this angular separation, each central maximum in the two intensity curves of Fig. 37-12 is centered on the first minimum of the other curve.

(b) What is the separation Δx of the centers of the *images* in the focal plane? (That is, what is the separation of the *central* peaks in the two curves?)

SOLUTION ■ The **Key Idea** here is to relate the separation Δx to the angle θ_i, which we now know. From either triangle between the lens and the screen in Fig. 37-12, we see that $\tan \theta_i/2 = \Delta x/2f$. Rearranging this and making the approximation $\tan \theta < \theta$, we find

$$\Delta x = f\theta_i, \tag{37-15}$$

where θ_i is in radian measure. Substituting known data then yields

$$\Delta x = (0.24\,\text{m})(2.1 \times 10^{-5}\,\text{rad}) = 5.0\,\mu\text{m.} \quad \text{(Answer)}$$

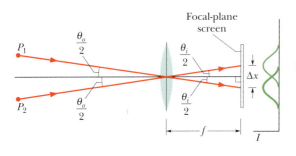

Focal-plane screen

FIGURE 37-12 ■ Light from two distant point objects P_1 and P_2 passes through a converging lens and forms images on a viewing screen in the focal plane of the lens. Only one representative ray from each object is shown. The images are not points but diffraction patterns, with intensities approximately as plotted at the right. The angular separation of the objects is θ_o and that of the images is θ_i; the central maxima of the images have a separation Δx.

37-6 Diffraction by a Double Slit

In the double-slit experiments of Chapter 36, we implicitly assumed that the slits were narrow compared to the wavelength of the light illuminating them; that is, $a \ll \lambda$. For such narrow slits, the central maximum of the diffraction pattern of either slit covers the entire viewing screen. Moreover, the interference of light from the two slits produces bright fringes that all have approximately the same intensity (Fig. 36-9).

In practice with visible light, however, the condition $a \ll \lambda$ is rarely met. For relatively wide slits, the interference of light from two slits produces bright fringes that do not all have the same intensity. That is, the intensities of the fringes produced by double-slit interference (as discussed in Chapter 36) are modified by diffraction of the light passing through each slit (as discussed in this chapter).

As an example, the intensity plot of Fig. 37-13a (like that in Fig. 36-9) suggests the double-slit interference pattern that would occur if the slits were infinitely narrow (for $a \ll \lambda$); all the bright interference fringes would have the same intensity. The intensity plot of Fig. 37-13b is that for diffraction by a single actual slit; the diffraction pattern has a broad central maximum and weaker secondary maxima at $\pm 1.7°$. The plot of Fig. 37-13c suggests the interference pattern for two actual slits. That plot was constructed by using the curve of Fig. 37-13b as an *envelope* on the intensity plot in Fig. 37-13a. The positions of the fringes are not changed; only the intensities are affected.

Figure 37-14a shows an actual pattern in which both double-slit interference and diffraction are evident. If one slit is covered, the single-slit diffraction pattern of Fig. 37-14b results. Note the correspondence between Figs. 37-14a and 37-13c and between Figs. 37-14b and 37-13b. In comparing these figures, bear in mind that 37-14 has been deliberately overexposed to bring out the faint secondary maxima and that two secondary maxima (rather than one) are shown.

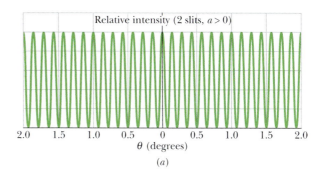

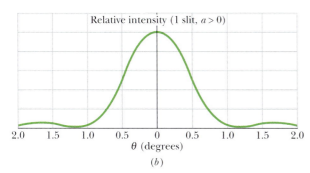

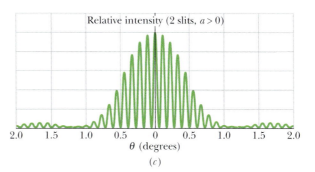

FIGURE 37-13 ■ (a) The intensity plot to be expected in a double-slit interference experiment with vanishingly narrow slits (here the distance between the center of the slits is $d = 25$ mm and the incident light is reddish-orange with $\lambda = 623$ mm). (b) The intensity plot for diffraction by a typical slit of width $a = 0.031$ mm (not vanishingly narrow). (c) The intensity plot to be expected for two slits of width $a = 0.031$ mm. The curve of (b) acts as an envelope, limiting the intensity of the double-slit fringes in (a). Note that the first minima of the diffraction pattern of (b) eliminate the double-slit fringes that would occur near 1.2° in (c).

With diffraction effects taken into account, the intensity of a double-slit interference pattern is given by

$$I(\theta) = I^{\max}(\cos^2 \beta)\left(\frac{\sin \alpha}{\alpha}\right)^2 \qquad \text{(double slit),} \qquad (37\text{-}16)$$

in which

$$\beta = \frac{\pi d}{\lambda} \sin \theta \qquad (37\text{-}17)$$

and

$$\alpha = \frac{\pi a}{\lambda} \sin \theta \qquad (37\text{-}18)$$

Here d is the distance between the centers of the slits, and a is the slit width. Note carefully that the right side of Eq. 37-16 is the product of $I^{\max}$ and two factors. (1) The

FIGURE 37-14 ■ (a) Interference fringes for an actual double-slit system; compare with Fig. 37-13c. (b) The diffraction pattern of a single slit; compare with Fig. 37-13b.

interference factor $\cos^2 \beta$ is due to the interference between two slits with slit separation d (as given by Eqs. 36-17 and 36-18). (2) The *diffraction factor* $[(\sin \alpha)/\alpha]^2$ is due to diffraction by a single slit of width a (as given by Eqs. 37-5 and 37-6).

Let us check these factors. If we let $a \rightarrow 0$ in Eq. 37-18, for example, then $\alpha \rightarrow 0$ and using L'Hopital's rule, we find that $(\sin \alpha)/\alpha \rightarrow 1$. Equation 37-16 then reduces, as it must, to an equation describing the interference pattern for a pair of vanishingly narrow slits with slit separation d. Similarly, putting $d = 0$ in Eq. 37-17 is equivalent physically to causing the two slits to merge into a single slit of width a. Then Eq. 37-17 yields $\beta = 0$ and $\cos^2 \beta = 1$. In this case Eq. 37-16 reduces, as it must, to an equation describing the diffraction pattern for a single slit of width a.

The double-slit pattern described by Eq. 37-16 and displayed in Fig. 37-14a combines interference and diffraction in an intimate way. Both are superposition effects, in that they result from the combining of waves with different phases at a given point. If the combining waves originate from a small number of elementary coherent sources—as in a double-slit experiment with $a \ll \lambda$—we call the process *interference*. If the combining waves originate in a single wavefront—as in a single-slit experiment—we call the process *diffraction*. This distinction between interference and diffraction (which is somewhat arbitrary and not always adhered to) is a convenient one, but we should not forget that both are superposition effects and usually both are present simultaneously (as in Fig. 37-14a).

TOUCHSTONE EXAMPLE 37-4: Bright Fringes

Let's consider a double slit with an unusually small spacing. Suppose the wavelength λ of the light source is 405 nm, the slit separation d is 19.44 μm, and the slit width a is 4.050 μm. Consider the interference of the light from the two slits and also the diffraction of the light through each slit.

(a) How many bright interference fringes are within the central peak of the diffraction envelope?

SOLUTION ■ Let us first analyze the two basic mechanisms responsible for the optical pattern produced in the experiment:

Single-slit diffraction: The **Key Idea** here is that the limits of the central peak are the first minima in the diffraction pattern due to either slit, individually. (See Fig. 37-13.) The angular locations of those minima are given by Eq. 37-3 ($a \sin \theta = m\lambda$). Let us write this equation as $a \sin \theta = m_1 \lambda$, with the subscript 1 referring to the one-slit diffraction. For the first minima in the diffraction pattern, we substitute $m_1 = 1$, obtaining

$$a \sin \theta = \lambda. \qquad (37\text{-}19)$$

Double-slit interference: The **Key Idea** here is that the angular locations of the bright fringes of the double-slit interference pattern are given by Eq. 36-14, which we can write as

$$d \sin \theta = m_2 \lambda, \qquad \text{for } m_2 = 1, 2, 3, \dots. \qquad (37\text{-}20)$$

Here the subscript 2 refers to the double-slit interference.

We can locate the first diffraction minimum within the double-slit fringe pattern by dividing Eq. 37-20 by Eq. 37-19 and solving for m_2. By doing so and then substituting the given data, we obtain

$$m_2 = \frac{d}{a} = \frac{19.44 \ \mu\text{m}}{4.050 \ \mu\text{m}} = 4.8.$$

This tells us that the bright interference fringe for $m_2 = 4$ fits into the central peak of the one-slit diffraction pattern, but the fringe for

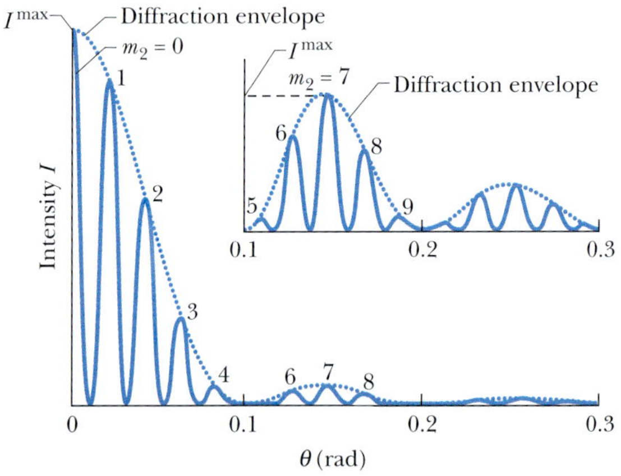

FIGURE 37-15 ■ One side of the intensity plot for a two-slit interference experiment; the diffraction envelope is indicated by the dotted curve. The smaller inset shows (vertically expanded) the intensity plot within the first and second side peaks of the diffraction envelope.

$m_2 = 5$ does not fit. Within the central diffraction peak we have the central bright fringe ($m_2 = 0$), and four bright fringes (up to $m_2 = 4$) on each side of it. Thus, a total of nine bright fringes of the double-slit interference pattern are within the central peak of the diffraction envelope. The bright fringes to one side of the central bright fringe are shown in Fig. 37-15.

(b) How many bright fringes are within either of the first side peaks of the diffraction envelope?

SOLUTION ▪ The **Key Idea** here is that the outer limits of the first side diffraction peaks are the second diffraction minima, each of which is at the angle θ given by $a \sin \theta = m_1 \lambda$ with $m_1 = 2$:

$$a \sin \theta = 2\lambda \qquad (37\text{-}21)$$

Dividing Eq. 37-20 by Eq. 37-21, we find

$$m_2 = \frac{2d}{a} = \frac{(2)(19.44\ \mu m)}{4.050\ \mu m} = 9.6.$$

This tells us that the second diffraction minimum occurs just before the bright interference fringe for $m_2 = 10$ in Eq. 37-20. Within either first side diffraction peak we have the fringes from $m_2 = 5$ to $m_2 = 9$ for a total of five bright fringes of the double-slit interference pattern (shown in the inset of Fig. 37-15). However, if the $m_2 = 5$ bright fringe, which is almost eliminated by the first diffraction minimum, is considered too dim to count, then only four bright fringes are in the first side diffraction peak.

37-7 Diffraction Gratings

One of the most useful tools in the study of light and of objects that emit and absorb light is the **diffraction grating.** A diffraction grating is a device that uses **interference** phenomena to seperate a beam of light by wavelength. A diffraction grating is a more elaborate form of the double-slit arrangement of Fig. 36-8. This device has a much greater number N of slits, often called *rulings*, perhaps as many as several thousand per millimeter. An idealized grating consisting of only five slits is represented in Fig. 37-16. When monochromatic light is sent through the slits, it forms narrow interference fringes that can be analyzed to determine the wavelength of the light. (Diffraction gratings can also be opaque surfaces with narrow parallel grooves arranged like the slits in Fig. 37-16. Light then scatters back from the grooves to form interference fringes rather than being transmitted through open slits.)

With monochromatic light incident on a diffraction grating, if we gradually increase the number of slits from two to a large number N, the intensity plot changes from the typical double-slit plot of Fig. 37-13c to a much more complicated one and then eventually to a simple graph like that shown in Fig. 37-17a. The pattern you would see on a viewing screen using monochromatic red light from, say, a helium-neon laser, is shown in Fig. 37-17b. The maxima are now very narrow (and so are called *lines*); they are separated by relatively wide dark regions.

We use a familiar procedure to find the locations of the bright lines on the viewing screen. We first assume that the screen is far enough from the grating so that the rays reaching a particular point P on the screen are approximately parallel when they leave the grating (Fig. 37-18). Then we apply to each pair of adjacent rulings the same reasoning we used for double-slit interference. The separation d between rulings is called the *grating spacing*. (If N rulings occupy a total width w, then $d = w/N$.) The path length difference between adjacent rays is again $d\sin\theta$ (Fig. 37-18), where θ is the angle from the central axis of the grating (and of the diffraction pattern) to point P. A line will be located at P if the path length difference between adjacent rays is an integer number of wavelengths—that is, if

$$d \sin \theta = m\lambda, \qquad \text{for } m = 0, 1, 2, \ldots \quad \text{(maxima—lines)}, \qquad (37\text{-}22)$$

where λ is the wavelength of the light. Each integer m represents a different line; hence these integers can be used to label the lines, as in Fig. 37-17. The integers are then called the *order numbers*, and the lines are called the zeroth-order line (the central line, with $m = 0$), the first-order line, the second-order line, and so on.

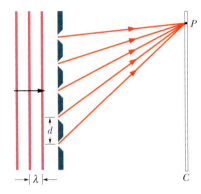

FIGURE 37-16 ▪ An idealized diffraction grating, consisting of only five rulings, that produces an interference pattern on a distant viewing screen C.

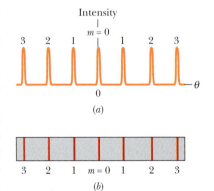

FIGURE 37-17 ▪ A diffraction grating illuminated with a single wavelength of light. (a) The intensity plot produced by a diffraction grating with a great many rulings consists of narrow peaks, here labeled with their order numbers m. (b) The corresponding bright fringes seen on the screen are called lines and are here also labeled with order numbers m. Lines of the zeroth, first, second, and third orders are shown.

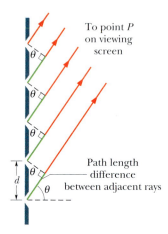

FIGURE 37-18 ■ The rays from the rulings in a diffraction grating to a distant point P are approximately parallel. The path length difference between each two adjacent rays is $d \sin\theta$, where θ is measured as shown. (The rulings extend into and out of the page.)

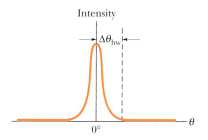

FIGURE 37-19 ■ The half-width $\Delta\theta_{hw}$ of the central line is measured from the center of that line to the adjacent minimum on a plot of I versus θ like Fig. 37-17a.

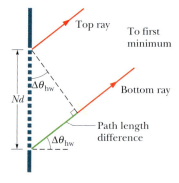

FIGURE 37-20 ■ The top and bottom rulings of a diffraction grating of N rulings are separated by distance Nd. The top and bottom rays passing through these rulings have a path length difference of $Nd \sin \Delta\theta_{hw}$, where $\Delta\theta_{hw}$ is the angle to the first minimum. (The angle is here greatly exaggerated for clarity.)

If we rewrite Eq. 37-22 as $\theta = \sin^{-1}(m\lambda/d)$ we see that, for a given diffraction grating, the angle from the central axis to any line (say, the third-order line) depends on the wavelength of the light being used. Thus, when light of an unknown wavelength is sent through a diffraction grating, measurements of the angles to the higher-order lines can be used in Eq. 37-22 to determine the wavelength. Even light of several unknown wavelengths can be distinguished and identified in this way. We cannot do that with the double-slit arrangement of Section 36-4, even though the same equation and wavelength dependence apply there. In double-slit interference, the bright fringes due to different wavelengths overlap too much to be distinguished.

Width of the Lines

A grating's ability to resolve (separate) lines of different wavelengths depends on the width of the lines. We shall here derive an expression for the *half-width* of the central line (the line for which $m = 0$) and then state an expression for the half-widths of the higher-order lines. We measure the half-width of the central line as the angle $\Delta\theta_{hw}$ from the center of the line at $\theta = 0$ outward to where the line effectively ends and darkness effectively begins with the first minimum (Fig. 37-19). At such a minimum, the N rays from the N slits of the grating cancel one another. (The actual width of the central line is, of course $2(\Delta\theta_{hw})$, but line widths are usually compared via half-widths.)

In Section 37-2 we were also concerned with the cancellation of a great many rays, there due to diffraction through a single slit. We obtained Eq. 37-3, which, because of the similarity of the two situations, we can use to find the first minimum here. It tells us that the first minimum occurs where the path length difference between the top and bottom rays equals λ. For single-slit diffraction, this difference is $a \sin \theta$. For a grating of N rulings, each separated from the next by distance d, the distance between the top and bottom rulings is Nd (Fig. 37-20), so the path length difference between the top and bottom rays here is $Nd \sin\Delta\theta_{hw}$. Thus, the first minimum occurs where

$$Nd \sin \Delta\theta_{hw} = \lambda. \tag{37-23}$$

Because $\Delta\theta_{hw}$ is small, $\sin\Delta\theta_{hw} \approx \Delta\theta_{hw}$ (in radian measure). Substituting this in Eq. 37-23 gives the half-width of the central line as

$$\Delta\theta_{hw} = \frac{\lambda}{Nd} \qquad \text{(half-width of central line)}. \tag{37-24}$$

We state without proof that the half-width of any other line depends on its location relative to the central axis and is

$$\Delta\theta_{hw} = \frac{\lambda}{Nd\cos\theta} \qquad \text{(half-width of line at θ)}. \tag{37-25}$$

Note that for light of a given wavelength λ and a given ruling separation d, the widths of the lines decrease with an increase in the number N of rulings. Thus, of two diffraction gratings, the grating with the larger value of N is better able to distinguish between wavelengths because its diffraction lines are narrower and so produce less overlap. But the line width of a monochromatic light beam is determined by the number of slits that the beam encounters. In a diffraction grating spectrometer, a collimating telescope can be used to illuminate all N slits of the grating.

The Diffraction Grating Spectrometer

Diffraction gratings are widely used to determine the wavelengths that are emitted by sources of light ranging from lamps to stars. Figure 37-21 shows a simple *grating spectroscope* in which a grating is used for this purpose. Light from source S is focused by lens L_1 on a vertical slit S_1 placed in the focal plane of lens L_2. The light emerging from tube C (called a *collimator*) is a plane wave and is incident perpendicularly on grating G, where it is diffracted into a diffraction pattern, with the $m = 0$ order diffracted at angle $\theta = 0$ along the central axis of the grating.

We can view the diffraction pattern that would appear on a viewing screen at any angle θ simply by orienting telescope T in Fig. 37-21 to that angle. Lens L_3 of the telescope then focuses the light diffracted at angle θ (and at slightly smaller and larger angles) onto a focal plane FF' within the telescope. When we look through eyepiece E, we see a magnified view of this focused image.

By changing the angle θ of the telescope, we can examine the entire diffraction pattern. For any order number other than $m = 0$, the original light is spread out according to wavelength (or color) so that we can determine, with Eq. 37-22, just what wavelengths are being emitted by the source. If the source emits a number of discrete wavelengths, what we see as we rotate the telescope horizontally through the angles corresponding to an order m is a vertical line of color for each wavelength, with the shorter-wavelength line at a smaller angle $m = 0$ than the longer-wavelength line.

For example, the light emitted by a hydrogen lamp, which contains hydrogen gas, has four discrete wavelengths in the visible range. If our eyes intercept this light directly, it appears to be white. If, instead, we view it through a grating spectroscope, we can distinguish, in several orders, the lines of the four colors corresponding to these visible wavelengths. (Such lines are called *emission lines*.) Four orders are represented in Fig. 37-22. In the central order ($m = 0$), the lines corresponding to all four wavelengths are superimposed, giving a single white line at $\theta = 0$. The colors are separated in the higher orders.

The third order is not shown in Fig. 37-22 for the sake of clarity; it actually overlaps the second and fourth orders. The fourth-order red line is missing because it is not formed by the grating used here. That is, when we attempt to solve Eq. 37-22 for

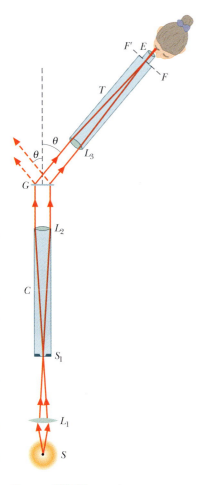

FIGURE 37-21 ■ A simple type of grating spectroscope used to analyze the wavelengths of light emitted by source S.

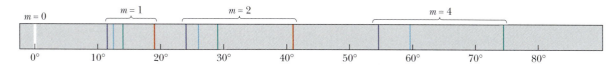

FIGURE 37-22 ■ The zeroth, first, second, and fourth orders of the visible emission lines from hydrogen. Note that the lines are farther apart at greater angles. (The lines are also dimmer and wider, although that is not shown here. Also, the third order line is eliminated for clarity.)

the angle θ for the red wavelength when $m = 4$, we find that $\sin \theta$ is greater than unity, which is not possible. The fourth order is then said to be *incomplete* for this grating; it might not be incomplete for a grating with greater spacing d, which will spread the lines less than in Fig. 37-22. Figure 37-23 is a photograph of the visible emission lines produced by cadmium.

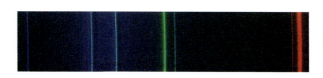

FIGURE 37-23 ■ The visible emission lines of cadmium, as seen through a grating spectroscope.

READING EXERCISE 37-5: The figure shows lines of different orders produced by a diffraction grating in monochromatic red light. (a) Is the center of the pattern to the left or right? (b) If we switch to monochromatic green light, will the half-widths of the lines then produced in the same orders be greater than, less than, or the same as the half-widths of the lines shown? ∎

37-8 Gratings: Dispersion and Resolving Power

There are two characteristics that are important in the design of a diffraction grating spectrometer. First, the different wavelengths of light in a beam should be spread out. This characteristic is called **dispersion.** The second characteristic is the **resolving power** of the spectrometer. It should have a narrow line width for each wavelength so the lines are sharp.

The fine rulings, each 0.5 μm wide, on a compact disc function as a diffraction grating. When a small source of white light illuminates a disc, the diffracted light forms colored "lanes" that are the composite of the diffraction patterns from the rulings.

Dispersion

To be useful in distinguishing wavelengths that are close to each other (as in a grating spectroscope), a grating must spread apart the diffraction lines associated with the various wavelengths. This spreading, called **dispersion,** is defined as

$$D = \frac{\Delta\theta}{\Delta\lambda} \qquad \text{(dispersion defined).} \qquad (37\text{-}26)$$

Here $\Delta\theta$ is the angular separation of two lines whose wavelengths differ by $\Delta\lambda$. The greater D is, the greater is the distance between two emission lines whose wavelengths differ by $\Delta\lambda$. We show below that the dispersion of a grating at angle θ is given by

$$D = \frac{m}{d\,\cos\theta} \qquad \text{(dispersion of a grating).} \qquad (37\text{-}27)$$

Thus, to achieve higher dispersion we must use a grating of smaller grating spacing d and work in a higher order m. Note that the dispersion does not depend on the number of rulings. The SI unit for D is the degree per meter or the radian per meter.

Proof of Eq. 37-27

Let us start with Eq. 37-22, the expression for the locations of the lines in the diffraction pattern of a grating:

$$d\,\sin\theta = m\lambda.$$

Let us regard θ and λ as variables and take differentials of this equation. We find

$$d\,\cos\theta\,(d\theta) = m\,(d\lambda),$$

where the differentials $d\theta$ and $d\lambda$ are placed in parentheses to distinguish them from the product of the center to center slit spacing d and the angle θ or wavelength λ.

For small enough angles, we can write these differentials as small differences, obtaining

$$d\,\cos\theta\,(\Delta\theta) = m(\Delta\lambda), \qquad (37\text{-}30)$$

or

$$\frac{(\Delta\theta)}{(\Delta\lambda)} = \frac{m}{d\,\cos\theta}.$$

The ratio on the left is simply D (see Eq. 37-26), so we have indeed derived Eq. 37-27.

Resolving Power

To *resolve* lines whose wavelengths are close together (that is, to make the lines distinguishable), the line should also be as narrow as possible. Expressed otherwise, the grating should have a high **resolving power R,** defined as

$$R = \frac{\langle \lambda \rangle}{\Delta \lambda} \qquad \text{(resolving power defined).} \qquad (37\text{-}28)$$

Here $\langle \lambda \rangle$ is the mean wavelength of two emission lines that can barely be recognized as separate, and $\Delta \lambda$ is the wavelength difference between them. The greater R is, the closer two emission lines can be and still be resolved. We shall show below that the resolving power of a grating is given by the simple expression

$$R = Nm \qquad \text{(resolving power of a grating).} \qquad (37\text{-}29)$$

To achieve high resolving power, we must spread out the light beam so it is incident on many rulings (large N in Eq. 37-29).

Proof of Eq. 37-29

We start with Eq. 37-30, which was derived from Eq. 37-22, the expression for the locations of the lines in the diffraction pattern formed by a grating. Here $\Delta \lambda$ is the small wavelength difference between two waves that are diffracted by the grating, and $\Delta \theta$ is the angular separation between them in the diffraction pattern. If $\Delta \theta$ is to be the smallest angle that will permit the two lines to be resolved, it must (by Rayleigh's criterion) be equal to the half-width of each line, which is given by Eq. 37-25:

$$\Delta \theta_{hw} = \frac{\lambda}{Nd \cos \theta}.$$

If we substitute $\Delta \theta_{hw}$ as given here for $\Delta \theta$ in Eq. 37-30, we find that

$$\frac{\lambda}{N} = m \Delta \lambda,$$

from which it readily follows that

$$R = \frac{\lambda}{\Delta \lambda} = Nm.$$

This is Eq. 37-29, which we set out to derive.

Dispersion and Resolving Power Compared

The resolving power of a grating must not be confused with its dispersion. Table 37-1 shows the characteristics of three gratings, all illuminated with light of wavelength $\lambda = 589$ nm, whose diffracted light is viewed in the first order ($m = 1$ in Eq. 37-22). You should verify that the values of D and R as given in the table can be calculated with Eqs. 37-27 and 37-29, respectively. (In the calculations for D, you will need to convert radians per meter to degrees per micrometer.)

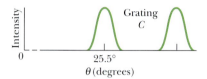

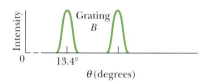

FIGURE 37-24: The intensity patterns for light of two wavelengths sent through the gratings of Table 37-1. Grating B has the highest resolving power and grating C the highest dispersion.

TABLE 37-1
Three Gratings^a

Grating	Specifications		Calculated Values		
	N	d (nm)	θ	D (°/μm)	R
A	10 000	2540	13.4°	23.2	10 000
B	20 000	2540	13.4°	23.2	20 000
C	10 000	1370	25.5°	46.3	10 000

^{a}Data are for $\lambda = 589$ nm and $m = 1$.

For the conditions noted in Table 37-1, gratings A and B have the same *dispersion* and A and C have the same *resolving power.*

Figure 37-24 shows the intensity patterns (also called *line shapes*) that would be produced by these gratings for two lines of wavelengths λ_1 and λ_2, in the vicinity of $\lambda = 589$ nm. Grating B, with the higher resolving power, produces narrower lines and thus is capable of distinguishing lines that are much closer together in wavelength than those in the figure. Grating C, with the higher dispersion, produces the greater angular separation between the lines.

TOUCHSTONE EXAMPLE 37-5: Diffraction Grating

A diffraction grating has 1.26×10^4 rulings uniformly spaced over width $w = 25.4$ mm (so that it has 496 lines/mm). It is illuminated at normal incidence by yellow light from a sodium vapor lamp. This light contains two closely spaced emission lines (known as the sodium doublet) of wavelengths 589.00 nm and 589.59 nm.

(a) At what angle does the first-order maximum occur (on either side of the center of the diffraction pattern) for the wavelength of 589.00 nm?

SOLUTION ■ The **Key Idea** here is that the maxima produced by the diffraction grating can be located with Eq. 37-22 ($d \sin \theta = m\lambda$). The grating spacing d for this diffraction grating is

$$d = \frac{w}{N} = \frac{25.4 \times 10^{-3} \text{ m}}{1.26 \times 10^4}$$

$$= 2.016 \times 10^{-6} \text{ m} = 2016 \text{ nm}.$$

The first-order maximum corresponds to $m = 1$. Substituting these values for d and m into Eq. 37-22 leads to

$$\theta = \sin^{-1}\frac{m\lambda}{d} = \sin^{-1}\frac{(1)(589.00 \text{ nm})}{2016 \text{ nm}}$$

$$= 16.99° \approx 17.0°. \qquad \text{(Answer)}$$

(b) Using the dispersion of the grating, calculate the angular separation between the two lines in the first order.

SOLUTION ■ One **Key Idea** here is that the angular separation $\Delta\theta$ between the two lines in the first order depends on their wavelength difference $\Delta\lambda$ and the dispersion D of the grating, according to Eq. 37-26 ($D = \Delta\theta/\Delta\lambda$). A second **Key Idea** is that the dispersion D depends on the angle θ at which it is to be evaluated. We can assume that, in the first order, the two sodium lines occur close enough to each other for us to evaluate D at the angle $\theta = 16.99°$ we found in part (a) for one of those lines. Then Eq. 37-27 gives the dispersion as

$$D = \frac{m}{d \cos\theta} = \frac{1}{(2016 \text{ nm})(\cos 16.99°)}$$

$$= 5.187 \times 10^{-4} \text{ rad/nm}.$$

From Eq. 37-26, we then have

$$\Delta\theta = D \, \Delta\lambda = (5.187 \times 10^{-4} \text{ rad/nm})(589.59 \text{ nm} - 589.00 \text{ nm})$$

$$= 3.06 \times 10^{-4} \text{ rad} = 0.0175°. \qquad \text{(Answer)}$$

You can show that this result depends on the grating spacing d but not on the number of rulings there are in the grating.

(c) What is the least number of rulings a grating can have and still be able to resolve the sodium doublet in the first order?

SOLUTION ■ One **Key Idea** here is that the resolving power of a grating in any order m is physically set by the number of rulings N in the grating according to Eq. 37-29 ($R = Nm$). A second **Key Idea** is that the least wavelength difference $\Delta\lambda$ that can be resolved depends on the average wavelength involved and the resolving power R of the grating, according to Eq. 37-28 ($R = \langle\lambda\rangle/\Delta\lambda$).

For the sodium doublet to be barely resolved, $\Delta\lambda$ must be their wavelength separation of 0.59 nm, and $\langle\lambda\rangle$ must be their average wavelength of 589.30 nm.

Putting these ideas together, we find that the least number of rulings for a grating to resolve the sodium doublet is

$$N = \frac{R}{m} = \frac{\langle\lambda\rangle}{m\Delta\lambda}$$

$$= \frac{589.30 \text{ nm}}{(1)(0.59 \text{ nm})} = 999 \text{ rulings.} \qquad \text{(Answer)}$$

37-9 X-Ray Diffraction

X rays are electromagnetic radiation whose wavelengths are of the order of 1 Å ($= 0.1$ nm $= 10^{-10}$ m). Compare this with a wavelength of 550 nm ($= 5.5 \times 10^{-7}$ m) at the center of the visible spectrum. Figure 37-25 shows that x rays are produced when electrons escaping from a heated filament F are accelerated by a potential difference V and strike a metal target T.

A standard optical diffraction grating cannot be used to discriminate between different wavelengths in the x-ray wavelength range. For $\lambda = 1$ Å ($= 0.1$ nm) and $d = 3000$ nm, for example, Eq. 37-22 shows that the first-order maximum occurs at

$$\theta = \sin^{-1}\frac{m\lambda}{d} = \sin^{-1}\frac{(1)(0.1 \text{ nm})}{3000 \text{ nm}} = 0.0019°.$$

This is too close to the central maximum to be practical. A grating with $d \approx \lambda$ is desirable, but, since x-ray wavelengths are about equal to atomic diameters, such gratings cannot be constructed mechanically.

In 1912, it occurred to German physicist Max von Laue that a crystalline solid, which consists of a regular array of atoms, might form a natural three-dimensional "diffraction grating" for x rays. The idea is that, in a crystal such as sodium chloride (NaCl), a basic unit of atoms (called the *unit cell*) repeats itself throughout the array. In NaCl four sodium ions and four chlorine ions are associated with each unit cell. Figure 37-26a represents a section through a crystal of NaCl and identifies this basic unit. The unit cell is a cube measuring a_0 on each side.

When an x-ray beam enters a crystal such as NaCl, x rays are *scattered*—that is, redirected—in all directions by the crystal structure. In some directions the scattered

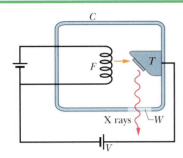

FIGURE 37-25: X rays are generated when electrons leaving heated filament F are accelerated through a potential difference V and strike a metal target T. The "window" W in the evacuated chamber C is transparent to x rays.

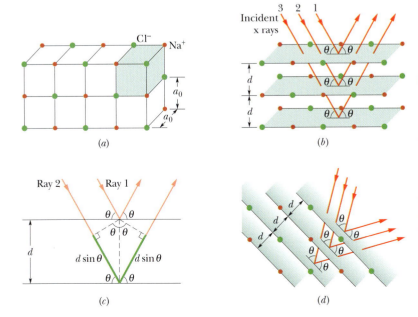

(a)

(c)

(b)

(d)

FIGURE 37-26: (a) The cubic structure of NaCl, showing the sodium and chlorine ions and a unit cell (shaded). (b) Incident x rays undergo diffraction by the structure of (a). The x rays are diffracted as if they were reflected by a family of parallel planes, with the angle of reflection equal to the angle of incidence, both angles measured relative to the planes (not relative to a normal as in optics). (c) The path length difference between waves effectively reflected by two adjacent planes is $2d\sin\theta$. (d) A different orientation of the incident x rays relative to the structure. A different family of parallel planes now effectively reflects the x rays.

waves undergo destructive interference, resulting in intensity minima; in other directions the interference is constructive, resulting in intensity maxima. This process of scattering and interference is a form of diffraction, although it is unlike the diffraction of light traveling through a slit or past an edge as we discussed earlier.

Although the process of diffraction of x rays by a crystal is complicated, the maxima turn out to be in directions as if the x rays were reflected by a family of parallel *reflecting planes* (or *crystal planes*) that extend through the atoms within the crystal and that contain regular arrays of the atoms. (The x rays are not actually reflected; we use these fictional planes only to simplify the analysis of the actual diffraction process.)

Figure 37-26b shows three of the family of planes, with *interplanar spacing d*, from which the incident rays shown are said to reflect. Rays 1, 2, and 3 reflect from the first, second, and third planes, respectively. At each reflection the angle of incidence and the angle of reflection are represented with θ. Contrary to the custom in optics, these angles are defined relative to the *surface* of the reflecting plane rather than a normal to that surface. For the situation of Fig. 37-26b, the interplanar spacing happens to be equal to the unit cell dimension a_0.

Figure 37-26c shows an edge-on view of reflection from an adjacent pair of planes. The waves of rays 1 and 2 arrive at the crystal in phase. After they are reflected, they must again be in phase, because the reflections and the reflecting planes have been defined solely to explain the intensity maxima in the diffraction of x rays by a crystal. Unlike light rays, the x rays have negligible refraction when entering the crystal; moreover, we do not define an index of refraction for this situation. Thus, the relative phase between the waves of rays 1 and 2 as they leave the crystal is set solely by their path length difference. For these rays to be in phase, the path length difference must be equal to an integer multiple of the wavelength λ of the x rays.

By drawing the dashed perpendiculars in Fig. 37-26c, we find that the path length difference is $2d \sin \theta$. In fact, this is true for any pair of adjacent planes in the family of planes represented in Fig. 37-26b. Thus, we have, as the criterion for intensity maxima for x-ray diffraction,

$$2d \sin\theta = m\lambda, \qquad \text{for } m = 1, 2, 3, \ldots \quad \text{(Bragg's law),} \qquad (37\text{-}31)$$

where m is the order number of an intensity maximum. Equation 37-31 is called **Bragg's law** after British physicist W. L. Bragg, who first derived it. (He and his father shared the 1915 Nobel Prize for their use of x rays to study the structures of crystals.) The angle of incidence and reflection in Eq. 37-31 is called a *Bragg angle*.

Regardless of the angle at which x rays enter a crystal, there is always a family of planes from which they can be said to reflect so that we can apply Bragg's law. In Fig. 37-26d, the crystal structure has the same orientation as it does in Fig. 37-26a, but the angle at which the beam enters the structure differs from that shown in Fig. 37-26b. This new angle requires a new family of reflecting planes, with a different interplanar spacing d and different Bragg angle θ, in order to explain the x-ray diffraction via Bragg's law.

Figure 37-27 shows how the interplanar spacing d can be related to the unit cell dimension a_0. For the particular family of planes shown there, the Pythagorean theorem gives

$$5d = \sqrt{5}a_0,$$

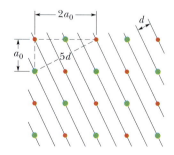

FIGURE 37-27: A family of planes through the structure of Fig. 37-26a, and a way to relate the edge length a_0 of a unit cell to the interplanar spacing d.

or

$$d = \frac{a_0}{\sqrt{5}}. \qquad (37\text{-}32)$$

Figure 37-27 suggests how the dimensions of the unit cell can be found once the interplanar spacing has been measured by means of x-ray diffraction.

X-ray diffraction is a powerful tool for studying both x-ray spectra and the arrangement of atoms in crystals. To study spectra, a particular set of crystal planes, having a known spacing d, is chosen. These planes effectively reflect different wavelengths at different angles. A detector that can discriminate one angle from another can then be used to determine the wavelength of radiation reaching it. The crystal itself can be studied with a monochromatic x-ray beam, to determine not only the spacing of various crystal planes but also the structure of the unit cell.

Problems

SEC. 37-2 ■ DIFFRACTION BY A SINGLE SLIT: LOCATING THE MINIMA

1. Narrow Slit Light of wavelength 633 nm is incident on a narrow slit. The angle between the first diffraction minimum on one side of the central maximum and the first minimum on the other side is 1.20°. What is the width of the slit?

2. Distance Between Monochromatic light of wavelength 441 nm is incident on a narrow slit. On a screen 2.00 m away, the distance between the second diffraction minimum and the central maximum is 1.50 cm. (a) Calculate the angle of diffraction θ of the second minimum. (b) Find the width of the slit.

3. Single Slit A single slit is illuminated by light of wavelengths λ_a and λ_b, chosen so the first diffraction minimum of the λ_a component coincides with the second minimum of the λ_b component. (a) What relationship exists between the two wavelengths? (b) Do any other minima in the two diffraction patterns coincide?

4. First and Fifth The distance between the first and fifth minima of a single-slit diffraction pattern is 0.35 mm with the screen 40 cm away from the slit, when light of wavelength 550 nm is used. (a) Find the slit width. (b) Calculate the angle θ of the first diffraction minimum.

5. Plane Wave A plane wave of wavelength 590 nm is incident on a slit with a width of $a = 0.40$ nm. A thin converging lens of focal length +70 cm is placed between the slit and a viewing screen and focuses the light on the screen. (a) How far is the screen from the lens? (b) What is the distance on the screen from the center of the diffraction pattern to the first minimum?

6. Sound Waves Sound waves with frequency 3000 Hz and speed 343 m/s diffract through the rectangular opening of a speaker cabinet and into a large auditorium. The opening, which has a horizontal width of 30.0 cm, faces a wall 100 m away (Fig. 37-28). Where along that wall will a listener be at the first diffraction minimum and thus have difficulty hearing the sound? (Neglect reflections).

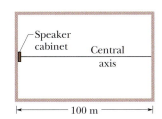

FIGURE 37-28 ■
Problem 6.

7. Central Maximum A slit 1.00 mm wide is illuminated by light of wavelength 589 nm. We see a diffraction pattern on a screen 3.00 m away. What is the distance between the first two diffraction minima on the same side of the central diffraction maximum?

SEC. 37-4 ■ INTENSITY IN SINGLE-SLIT DIFFRACTION, QUANTITATIVELY

8. Off Central Axis A 0.10-mm-wide slit is illuminated by light of wavelength 589 nm. Consider a point P on a viewing screen on which the diffraction pattern of the slit is viewed; the point is at 30° from the central axis of the slit. What is the phase difference between the Huygens wavelets arriving at point P from the top and midpoint of the slit? (*Hint:* See Eq. 37-4.)

9. Explain Quantitatively If you double the width of a single slit, the intensity of the central maximum of the diffraction pattern increases by a factor of 4, even though the energy passing through the slit only doubles. Explain this quantitatively.

10. Monochromatic Monochromatic light with wavelength 538 nm is incident on a slit with width 0.025 mm. The distance from the slit to a screen is 3.5 m. Consider a point on the screen 1.1 cm from the central maximum. (a) Calculate θ for that point. (b) Calculate α. (c) Calculate the ratio of the intensity at this point to the intensity at the central maximum.

11. FWHM The full width at half-maximum (FWHM) of a central diffraction maximum is defined as the angle between the two points in the pattern where the intensity is one-half that at the center of the pattern. (See Fig. 37-7b.) (a) Show that the intensity drops to one-half the maximum value when $\sin^2 \alpha = \alpha^2/2$. (b) Verify that $\alpha = 1.39$ rad (about 80°) is a solution to the transcendental equation of (a). (c) Show that the FWHM is $\Delta\theta = 2\sin^{-1}(0.443\lambda/a)$, where a is the slit width. (d) Calculate the FWHM of the central maximum for slits whose widths are 1.0, 5.0, and 10 wavelengths.

12. Babinet's Principle A monochromatic beam of parallel light is incident on a "collimating" hole of diameter $x \gg \lambda$. Point P lies in the geometrical shadow region on a *distant* screen (Fig. 37-29a). Two diffracting objects, shown in Fig. 37-29b, are placed in turn

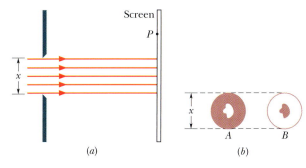

FIGURE 37-29 ■ Problem 12.

over the collimating hole. *A* is an opaque circle with a hole in it and *B* is the "photographic negative" of *A*. Using superposition concepts, show that the intensity at *P* is identical for the two diffracting objects *A* and *B*.

13. Values of α (a) Show that the values of α at which intensity maxima for single-slit diffraction occur can be found exactly by differentiating Eq. 37-5 with respect to α and equating the result to zero, obtaining the condition $\tan\alpha = \alpha$. (b) Find the values of α satisfying this relation by plotting the curve $y = \tan\alpha$ and the straight line $y = \alpha$ and finding their intersections or by using a calculator with an equation solver to find an appropriate value of α (or by using trial and error). (c) Find the (noninteger) values of m corresponding to successive maxima in the single-slit pattern. Note that the secondary maxima do not lie exactly halfway between minima.

SEC. 37-5 ■ DIFFRACTION BY A CIRCULAR APERTURE

14. Entopic Halos At night many people see rings (called *entopic halos*) surrounding bright outdoor lamps in otherwise dark surroundings. The rings are the first of the side maxima in diffraction patterns produced by structures that are thought to be within the cornea (or possibly the lens) of the observer's eye. (The central maxima of such patterns overlap the lamp.) (a) Would a particular ring become smaller or larger if the lamp were switched from blue to red light? (b) If a lamp emits white light, is blue or red on the outside edge of the ring? (c) Assume that the lamp emits light at wavelength 550 nm. If a ring has an angular diameter of 2.5°, approximately what is the (linear) diameter of the structure in the eye that causes the ring?

15. Headlights The two headlights of an approaching automobile are 1.4 m apart. At what (a) angular separation and (b) maximum distance will the eye resolve them? Assume that the pupil diameter is 5.0 mm, and use a wavelength of 550 nm for the light. Also assume that diffraction effects alone limit the resolution so that Rayleigh's criterion can be applied.

16. An Astronaut An astronaut in a space shuttle claims she can just barely resolve two point sources on the Earth's surface, 160 km below. Calculate their (a) angular and (b) linear separation, assuming ideal conditions. Take $\lambda = 540$ nm and the pupil diameter of the astronaut's eye to be 5.0 mm.

17. Moon's Surface Find the separation of two points on the Moon's surface that can just be resolved by the 200 in. (= 5.1 m) telescope at Mount Palomar, assuming that this separation is determined by diffraction effects. The distance from the Earth to the Moon is 3.8×10^5 km. Assume a wavelength of 550 nm for the light.

18. Large Room The wall of a large room is covered with acoustic tile in which small holes are drilled 5.0 mm from center to center. How far can a person be from such a tile and still distinguish the individual holes, assuming ideal conditions, the pupil diameter of the observer's eye to be 4.0 mm, and the wavelength of the room light to be 550 nm?

19. Estimate Linear Separation Estimate the linear separation of two objects on the planet Mars that can just be resolved under ideal conditions by an observer on Earth (a) using the naked eye and (b) using the 200 in. (= 5.1 m) Mount Palomar telescope. Use the following data: distance to Mars = 8.0×10^7 km, diameter of pupil = 5.0 mm, wavelength of light = 550 nm.

20. Radar System The radar system of a navy cruiser transmits at a wavelength of 1.6 cm, from a circular antenna with a diameter of 2.3 m. At a range of 6.2 km, what is the smallest distance that two speedboats can be from each other and still be resolved as two separate objects by the radar system?

21. Tiger Beetles The wings of tiger beetles (Fig. 37-30) are colored by interference due to thin cuticle-like layers. In addition, these layers are arranged in patches that are 60 μm across and produce different colors. The color you see is a pointillistic mixture of thin-film interference colors that varies with perspective. Approximately what viewing distance from a wing puts you at the limit of resolving the different colored patches according to Rayleigh's criterion? Use 550 nm as the wavelength of light and 3.00 mm as the diameter of your pupil.

FIGURE 37-30 ■ Problem 21. Tiger beetles are colored by pointillistic mixtures of thin-film interference colors.

22. Discovery In June 1985, a laser beam was sent out from the Air Force Optical Station on Maui, Hawaii, and reflected back from the shuttle *Discovery* as it sped by, 354 km overhead. The diameter of the central maximum of the beam at the shuttle position was said to be 9.1 m, and the beam wavelength was 500 nm. What is the effective diameter of the laser aperture at the Maui ground station? (*Hint:* A laser beam spreads only because of diffraction; assume a circular exit aperture.)

23. Millimeter-Wave Radar Millimeter-wave radar generates a narrower beam than conventional microwave radar, making it less vulnerable to antiradar missiles. (a) Calculate the angular width of the central maximum, from first minimum to first minimum, produced by a 220 GHz radar beam emitted by a 55.0-cm-diameter circular antenna. (The frequency is chosen to coincide with a low-absorption atmospheric "window.") (b) Calculate the same quantity for the ship's radar described in Problem 20.

24. Circular Obstacle A circular obstacle produces the same diffraction pattern as a circular hole of the same diameter (except very near $\theta = 0$). Airborne water drops are examples of such obstacles. When you see the Moon through suspended water drops, such as in a fog, you intercept the diffraction pattern from many drops.

FIGURE 37-31 ■ Problem 24. The corona around the Moon is a composite of the diffraction patterns of airborne water drops.

The composite of the central diffraction maxima of those drops forms a white region that surrounds the Moon and may obscure it. Figure 37-31 is a photograph in which the Moon is obscured. There are two, faint, colored rings around the Moon (the larger one may be too faint to be seen in your copy of the photograph). The smaller ring is on the outer edge of the central maxima from the drops; the somewhat larger ring is on the outer edge of the smallest of the secondary maxima from the drops (see Fig. 37-3). The color is visible because the rings are adjacent to the diffraction minima (dark rings) in the patterns. (Colors in other parts of the pattern overlap too much to be visible.)

(a) What is the color of these rings on the outer edges of the diffraction maxima? (b) The colored ring around the central maxima in Fig. 37-31 has an angular diameter that is 1.35 times the angular diameter of the Moon, which is 0.50°. Assume that the drops all have about the same diameter. Approximately what is that diameter?

25. Allegheny Observatory (a) What is the angular separation of two stars if their images are barely resolved by the Thaw refracting telescope at the Allegheny Observatory in Pittsburgh? The lens diameter is 76 cm and its focal length is 14 m. Assume $\lambda = 550$ nm. (b) Find the distance between these barely resolved stars if each of them is 10 light-years distant from Earth. (c) For the image of a single star in this telescope, find the diameter of the first dark ring in the diffraction pattern, as measured on a photographic plate placed at the focal plane of the telescope lens. Assume that the structure of the image is associated entirely with diffraction at the lens aperture and not with lens "errors".

26. Soviet–French Experiment In a joint Soviet–French experiment to monitor the Moon's surface with a light beam, pulsed radiation from a ruby laser ($\lambda = 0.69$ μm) was directed to the Moon through a reflecting telescope with a mirror radius of 1.3 m. A reflector on the Moon behaved like a circular plane mirror with radius 10 cm, reflecting the light directly back toward the telescope on the Earth. The reflected light was then detected after being brought

to a focus by this telescope. What fraction of the original light energy was picked up by the detector? Assume that for each direction of travel all the energy is in the central diffraction peak.

SEC. 37-6 ■ DIFFRACTION BY A DOUBLE SLIT

27. Bright Fringes Suppose that the central diffraction envelope of a double-slit diffraction pattern contains 11 bright fringes and the first diffraction minima eliminate (are coincident with) bright fringes. How many bright fringes lie between the first and second minima of the diffraction envelope?

28. Slit Separation In a double-slit experiment, the slit separation d is 2.00 times the slit width w. How many bright interference fringes are in the central diffraction envelope?

29. Eliminate Bright Fringes (a) In a double-slit experiment, what ratio of d to a causes diffraction to eliminate the fourth bright side fringe? (b) What other bright fringes are also eliminated?

30. Two Slits Two slits of width a and separation d are illuminated by a coherent beam of light of wavelength λ. What is the linear separation of the bright interference fringes observed on a screen that is at a distance D away?

31. How Many (a) How many bright fringes appear between the first diffraction-envelope minima to either side of the central maximum in a double-slit pattern if $\lambda = 550$ nm, $d = 0.150$ mm, and $a = 30.0$ μm? (b) What is the ratio of the intensity of the third bright fringe to the intensity of the central fringe?

32. Intensity Vs. Position Light of wavelength 440 nm passes through a double slit, yielding a diffraction pattern whose graph of intensity I versus angular position θ is shown in Fig. 37-32. Calculate the (a) slit width and (b) slit separation. (c) Verify the displayed intensities of the $m = 1$ and $m = 2$ interference fringes.

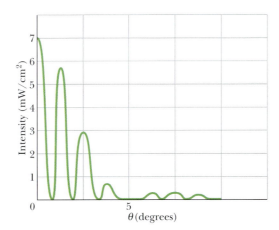

FIGURE 37-32 ■ Problem 32.

SEC. 37-7 ■ DIFFRACTION GRATINGS

33. Calculate d A diffraction grating 20.0 mm wide has 6000 rulings. (a) Calculate the distance d between adjacent rulings. (b) At what angles θ will intensity maxima occur on a viewing screen if the radiation incident on the grating has a wavelength of 589 nm?

34. Visible Spectrum A grating has 315 rulings/mm. For what wavelengths in the visible spectrum can fifth-order diffraction be observed when this grating is used in a diffraction experiment?

35. How Many Orders A grating has 400 lines/mm. How many orders of the entire visible spectrum (400–700 nm) can it produce in a diffraction experiment, in addition to the $m = 0$ order?

36. Confuse a Predator Perhaps to confuse a predator, some tropical gyrinid beetles (whirligig beetles) are colored by optical interference that is due to scales whose alignment forms a diffraction grating (which scatters light instead of transmiting it). When the incident light rays are perpendicular to the grating, the angle between the first-order maxima (on opposite sides of the zeroth-order maximum) is about 26° in light with a wavelength of 550 nm. What is the grating spacing of the beetle?

37. Two Adjacent Maxima Light of wavelength 600 nm is incident normally on a diffraction grating. Two adjacent maxima occur at angles given by $\sin \theta = 0.2$ and $\sin \theta = 0.3$. The fourth-order maxima are missing. (a) What is the separation between adjacent slits? (b) What is the smallest slit width this grating can have? (c) Which orders of intensity maxima are produced by the grating, assuming the values derived in (a) and (b)?

38. Normal Incidence A diffraction grating is made up of slits of width 300 nm with separation 900 nm. The grating is illuminated by monochromatic plane waves of wavelength $\lambda = 600$ nm at normal incidence. (a) How many maxima are there in the full diffraction pattern? (b) What is the width of a spectral line observed in the first order if the grating has 1000 slits?

39. Visible Spectrum Assume that the limits of the visible spectrum are arbitrarily chosen as 430 and 680 nm. Calculate the number of rulings per millimeter of a grating that will spread the first-order spectrum through an angle of 20°.

40. Gaseous Discharge Tube With light from a gaseous discharge tube incident normally on a grating with slit separation 1.73 μm, sharp maxima of green light are produced at angles $\theta = \pm\, 17.6°,\ 37.3°,\ -37.1°,\ 65.2°,$ and $-65.0°$. Compute the wavelength of the green light that best fits these data.

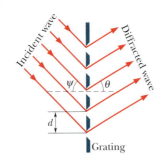

41. Show That Light is incident on a grating at an angle ψ as shown in Fig. 37-33. Show that bright fringes occur at angles θ that satisfy the equation

$$d(\sin \psi + \sin \theta) = m\lambda, \quad \text{for } m = 0, 1, 2, \ldots .$$

FIGURE 37-33 ■
Problem 41.

(Compare this equation with Eq. 37-22.) Only the special case $\psi = 0$ has been treated in this chapter.

42. Plot A grating with $d = 1.50$ μm is illuminated at various angles of incidence by light of wavelength 600 nm. Plot, as a function of the angle of incidence (0 to 90°), the angular deviation of the first-order maximum from the incident direction. (See Problem 41.)

43. Derive Derive Eq. 37-25, the expression for the half-widths of lines in a grating's diffraction pattern.

44. Spectrum Is Formed A grating has 350 rulings per millimeter and is illuminated at normal incidence by white light. A spectrum is formed on a screen 30 cm from the grating. If a hole 10 mm square is cut in the screen, its inner edge being 50 mm from the central

maximum and parallel to it, what is the range in the wavelengths of the light that passes through the hole?

45. Derive Two Derive this expression for the intensity pattern for a three-slit grating (ignore diffraction effects);

$$I_\theta = \tfrac{1}{9}I^{\max}(1 + 4 \cos\phi + 4 \cos^2 \phi),$$

where $\phi = (2\pi d \sin\theta)/\lambda$. Assume that $a \ll \lambda$; be guided by the derivation of the corresponding double-slit formula (Eq. 36-21).

SEC. 37-8 ■ GRATINGS: DISPERSION AND RESOLVING POWER

46. D Line The D line in the spectrum of sodium is a doublet with wave-lengths 589.0 and 589.6 nm. Calculate the minimum number of lines needed in a grating that will resolve this doublet in the second-order spectrum. See Touchstone Example 37-5.

47. Hydrogen–Deuterium Mix A source containing a mixture of hydrogen and deuterium atoms emits red light at two wavelengths whose mean is 656.3 nm and whose separation is 0.180 nm. Find the minimum number of lines needed in a diffraction grating that can resolve these lines in the first order.

48. Smallest Wavelength A grating has 600 rulings/mm and is 5.0 mm wide. (a) What is the smallest wavelength interval it can resolve in the third order at $\lambda = 500$ nm? (b) How many higher orders of maxima can be seen?

49. Dispersion Show that the dispersion of a grating is $D = (\tan \theta)/\lambda$.

50. Sodium Doublet With a particular grating the sodium doublet (see Touchstone Example 37-5) is viewed in the third order at 10° to the normal and is barely resolved. Find (a) the grating spacing and (b) the total width of the rulings.

51. Resolving Power A diffraction grating has resolving power $R = \langle\lambda\rangle/\Delta\,\lambda\ = Nm$. (a) Show that the corresponding frequency range Δf that can just be resolved is given by $\Delta f = c/Nm\lambda$. (b) From Fig. 37-18, show that the times required for light to travel along the ray at the bottom of the figure and the ray at the top differ by an amount $\Delta t = (Nd/c) \sin\theta$. (c) Show that $(\Delta f)(\Delta t) = 1$, this relation being independent of the various grating parameters. Assume $N \gg 1$.

52. Product (a) In terms of the angle θ locating a line produced by a grating, find the product of that line's half-width and the resolving power of grating. (b) Evaluate that product for the grating of Problem 38, for the first order.

SEC. 37-9 ■ X-RAY DIFFRACTION

53. Second-Order Reflection X rays of wavelength 0.12 nm are found to undergo second-order reflection at a Bragg angle of 28° from a lithium fluoride crystal. What is the interplanar spacing of the reflecting planes in the crystal?

54. Diffraction by Crystal Figure 37-34 is a graph of intensity versus angular position θ for the diffraction of an x-ray beam by a crystal. The beam consists of two wavelengths, and the spacing between the reflecting planes is 0.94 nm. What are the two wavelengths?

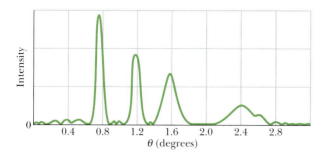

FIGURE 37-34 ■ Problem 54.

55. NaCl Crystal An x-ray beam of a certain wavelength is incident on a NaCl crystal, at 30.0° to a certain family of reflecting planes of spacing 39.8 pm. If the reflection from those planes is of the first order, what is the wavelength of the x rays?

56. Two Beams An x-ray beam of wavelength A undergoes first-order reflection from a crystal when its angle of incidence to a crystal face is 23°, and an x-ray beam of wavelength 97 pm undergoes third-order reflection when its angle of incidence to that face is 60°. Assuming that the two beams reflect from the same family of reflecting planes, find the (a) interplanar spacing and (b) wavelength A.

57. Not Possible Prove that it is not possible to determine both wavelength of incident radiation and spacing of reflecting planes in a crystal by measuring the Bragg angles for several orders.

58. Reflection Planes In Fig. 37-35, first-order reflection from the reflection planes shown occurs when an x-ray beam of wavelength

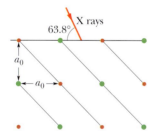

FIGURE 37-35 ■ Problem 58.

0.260 nm makes an angle of 63.8° with the top face of the crystal. What is the unit cell size a_0?

59. Square Crystal Consider a two-dimensional square crystal structure, such as one side of the structure shown in Fig. 37-26a. One interplanar spacing of reflecting planes is the unit cell size a_0. (a) Calculate and sketch the next five smaller interplanar spacings. (b) Show that your results in (a) are consistent with the general formula

$$d = \frac{a_0}{h^2 + k^2},$$

where h and k are relatively prime integers (they have no common factor other than unity).

60. X-Ray Beam In Fig. 37-36, an x-ray beam of wavelengths from 95.0 pm to 140 pm is incident at 45° to a family of reflecting planes with spacing $d = 275$ pm. At which wavelengths will these planes produce intensity maxima in their reflections?

61. NaCl In Fig. 37-36, let a beam of x-rays of wavelength 0.125 nm be incident on an NaCl crystal at an angle of 45.0° to the top face of the crystal and a family of reflecting planes. Let the reflecting planes have separation $d = 0.252$ nm. Through what angles must the crystal be turned about an axis that is perpendicular to the plane of the page for these reflecting planes to give intensity maxima in their reflections?

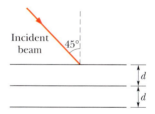

FIGURE 37-36 ■ Problems 60 and 61.

Additional Problems

62. Changing Interference Consider a plane wave of monochromatic green light, $\lambda = 500$ nm, that is incident normally upon two identical narrow slits (the widths of the individual slits are much less than λ). The slits are separated by a distance $d = 30$ μm. An interference pattern is observed on a screen located a distance L away from the slits. On the screen, the location nearest the central maximum where the intensity is zero (i.e., the first dark fringe) is found to be 1.5 cm from this central point. Let this particular position on the screen be referred to as P_1. (a) Calculate the distance, L, to the screen. Show all work. (b) In each of the parts below, one change has been made to the problem above (in each case, all parameters not explicitly mentioned have the value or characteristics stated above). For each case, explain briefly whether the light intensity at location P_1 remains zero or not. If not, does P_1 become the location of a maximum constructive interference (bright) fringe? In each case, explain your reasoning.

(1) One of the two slits is made slightly narrower, so that the amount of light passing through it is less than that through the other.

(2) The wavelength is doubled so that $\lambda = 1000$ nm.

(3) The two slits are replaced by a single slit whose width is exactly 60 μm.

63. Hearing and Seeing Around a Corner We can make the observation that we can hear around corners (somewhat) but not see around corners. Estimate why this is so by considering a doorway and two kinds of waves passing through it: (1) a beam of red light ($\lambda = 660$ nm), and (2) a sound wave playing an "A" ($f = 440$ Hz). (See Fig. 37-37.) Treat these two waves as plane waves passing through a slit whose width equals the width of the door. (a) Find the angle that gives the position of the first dark diffraction fringe. (b) From that, assuming you are 2 m back from the door, estimate how far outside the door you could be and still detect the wave. (See the picture for a clarification. The distance x is desired.)

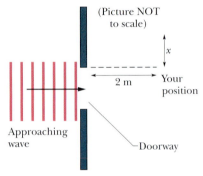

FIGURE 37-37 ■ Problem 63.

38 | Special Relativity

Guest Author: Edwin F. Taylor *Massachusetts Institute of Technology*

Billions of dollars have been spent constructing gigantic particle accelerators, such as this one 4 miles in circumference at Fermi National Accelerator Laboratory. More and more advanced accelerators give more and more energy and momentum to particles being accelerated. Decades of experimentation have verified that every particle, however great its energy and momentum, moves slower than the speed of light in a vacuum.

How can the energy of a particle increase without limit while its speed remains slower than the speed of light?

The answer is in this chapter.

38-1 Introduction

Special relativity and *general* relativity both describe the behavior of radiation and matter moving at or near the speed of light. However, special relativity is limited to situations in which gravitational effects can be neglected. Both special and general relativity are called *classical* theories because they do not describe atomic or molecular effects, for which quantum theory is needed.

Unfortunately, special relativity has a reputation for being difficult and mathematically complex. But if you understand basic algebra and square roots, you have the necessary mathematical tools to comprehend it. What makes special relativity seem difficult is that we have no direct experience with objects moving anywhere near the speed of light. It's no wonder that our idea of space and time, molded by everyday experience, is limited. As a result, the predictions of special relativity—fully verified by experiment—strike us as outlandish and outrageous. But these outlandish predictions not only make special relativity fascinating, they also provide us with deep insights into the nature of space and time—the arena in which we all live and in which science operates.

In this chapter we will show how the outlandish predictions of special relativity can be deduced logically from a single principle proposed by Albert Einstein at the beginning of the 20th century.

38-2 Origins of Special Relativity

While waiting at a stoplight, you notice that the car next to yours appears to be moving forward slowly (Fig. 38-1). Instead you suddenly realize that you are drifting backward, so you slam on the brakes to avoid bumping another car behind you. Before you step on the brakes, which car is standing still? Which is moving? Without seeing a "stationary" object such as a sign post, you cannot tell! Are such observations about relative motion trivial or profound? Can we cover the windows of our car and carry out some experiment inside—any experiment at all—to detect whether we are in motion or at rest?

Special relativity grew out of questions raised in the late 1800s and early 1900s about relative motions of material objects and waves. Some of these questions involved comparisons between everyday phenomena involving boats and ocean waves. Other questions were raised about the relative motions of objects and light waves. For example, consider ocean waves that move slowly past a swimmer moving in the same direction as the waves. The same waves will move rapidly past a second swimmer traveling in the opposite direction. Will the same thing happen when someone moves toward or away from light waves? Will observers traveling in opposite directions measure different speeds for the same light wave? Do light waves move in a *medium* the way ocean waves move in the medium of water?

From the age of 16 Albert Einstein (Fig. 38-2) puzzled over a thought experiment: Suppose you run very fast while looking at yourself in a mirror that you hold up in front of you. What happens as your running speed approaches the speed of light? Will the light waves move more and more slowly past you? In modern terms, *can you surf light waves?*

While Einstein was growing up, other people were trying to answer such questions with experiments. Some scientists hypothesized that light moves in a medium they called *ether*. In the late 1800s, Albert W. Michelson and Edward W. Morley carried out experiments with light trying to measure the motion of the Earth through this ether, under the assumption that the ether was at rest with respect to the Sun or some other location (such as the center of our galaxy). They used the fastest-moving object available to them: the Earth itself. The Earth moves around the sun at approximately 30

FIGURE 38-1 ■ While waiting at a stoplight, you find it hard to tell whether the car next to you is rolling forward or you are drifting backward—unless you are looking at a fixed object such as the speed limit sign.

FIGURE 38-2 ■ Albert Einstein in the early 1900s at the patent office in Bern, Switzerland, where he was employed when he published his article on special relativity. In later life he was known to dress much more informally.

kilometers per second in one direction in January and in the opposite direction past the sun in July. Michelson and Morley could detect no motion of the Earth through the hypothesized ether. These negative results caused great puzzlement.

In 1905 Einstein, a 26-year-old patent examiner in Bern, Switzerland, published a paper that changed the face of science.

READING EXERCISE 38-1: You are sitting in a train that stopped at a station ten minutes ago. Suddenly you notice that a second train on the track next to you is gliding past you. You feel a slight vibration that tells you your train is rolling slowly along the track. Is the second train in motion or at rest?   ■

38-3 The Principle of Relativity

Einstein's special relativity theory does not assume that light moves through a medium. Even so, it appears that Einstein did not base his ideas on Michelson and Morley's earlier failure to detect ether. Instead, Einstein treasured simplicity, logic, physical intuition, and his now famous thought experiments. He started from a clean assertion that he called the *Principle of Relativity*. Think of an automobile or train either at rest or moving at constant velocity. Define each of these enclosures as a *reference frame*. Then the *Principle of Relativity* says:

> All the laws of physics are the same in every reference frame.

In other words: Pull down the shades in your room or vehicle. Then carry out as many experiments as you need to create the laws of physics. Someone who carries out the same experiments inside another vehicle will discover the same laws, as long as this new vehicle moves at a constant velocity relative to yours.

The laws of physics contain fundamental numerical constants, such as the charge e on the electron, Planck's constant h, and the speed of light c in a vacuum. According to the Principle of Relativity, each of these constants must have the same numerical value when measured in any reference frame. In particular, all observers measure the speed of light in a vacuum to have the value presented back in Chapter 1 of $c = 299\ 792\ 458$ m/s ($\sim 3 \times 10^8$ m/s). The equality of the speed of light in all reference frames eliminates the need to postulate the existence of *ether* through which light propagates. The predictions of special relativity about space, time, mass, and motion all spring from the single *Principle of Relativity,* including the postulate of the "universal speed" of light.

The Principle of Relativity solved Einstein's puzzler about running fast while holding a mirror in front of you. You will not observe light waves to slow down as you move faster. Why not? Because, says the Principle of Relativity, light always moves past you with the same speed c, no matter how fast you run along the ground. *You cannot surf light waves!*

"Relativity theory" is a misleading term that Einstein avoided for years. What we call the special theory of relativity is based on the Principle of Relativity, which tells us that the laws of nature are the *same* for observers in different reference frames. These laws are *not* relative. *General* relativity employs an even more radical version than special relativity, of the Principle of Relativity—that the laws of nature are independent of the observer's viewpoint.

READING EXERCISE 38-2: While standing beside a railroad track, we are startled by a boxcar traveling past us at half the speed of light. A passenger (shown in the figure) standing at the front of the boxcar fires a laser pulse toward the rear of the boxcar. The pulse is

absorbed at the back of the boxcar. While standing beside the track we measure the speed of the pulse through the open side door. (a) Is our measured value of the speed of the pulse greater than, equal to, or less than its speed measured by the rider? (b) Is our measurement of the distance between emission and absorption of the light pulse greater than, equal to, or less than the distance between emission and absorption measured by the rider? (c) What conclusion can you draw about the relation between the times of flight of the light pulse as measured in the two reference frames? ∎

TOUCHSTONE EXAMPLE 38-1: Communications Storm!

A sunspot emits a tremendous burst of particles that travels toward the Earth. An astronomer on the Earth sees the emission through a solar telescope and issues a warning. The astronomer knows that when the particle pulse arrives it will wreak havoc with broadcast radio transmission. Communications systems require ten minutes to switch from over-the-air broadcast to underground cable transmission. What is the maximum speed of the particle pulse emitted by the Sun such that the switch can occur in time, between warning and arrival of the pulse? Take the sun to be 500 light-seconds distant from the Earth.

SOLUTION ∎ It takes 500 seconds for the warning light flash to travel the distance of 500 light-seconds between the Sun and the Earth and enter the astronomer's telescope. If the particle pulse moves at half the speed of light, it will take twice as long as light to reach the Earth. If the pulse moves at one-quarter the speed of light, it will take four times as long to make the trip. We generalize this by saying that if the pulse moves with speed v/c, it will take time Δt_{pulse} to make the trip given by the expression:

$$\Delta t_{pulse} = \frac{500 \text{ s}}{v_{pulse}/c}.$$

How long a warning time does the Earth astronomer have between arrival of the light flash carrying information about the pulse and the arrival of the pulse itself? It takes 500 seconds for the light to arrive. Therefore the warning time is the difference between the pulse transit time and the transit time of light:

$$\Delta t_{warning} = \Delta t_{pulse} - 500 \text{ s}.$$

But we know that the minimum possible warning time is 10 min = 600 s.

Therefore we have

$$600 \text{ s} = \frac{500 \text{ s}}{v_{pulse}/c} - 500 \text{ s},$$

which gives the maximum value for v_{pulse} if there is to be sufficient time for warning:

$$v_{pulse} = 0.455\,c. \qquad \text{(Answer)}$$

Observation reveals that pulses of particles emitted from the sun travel much slower than this maximum value. So we would have a much longer warning time than calculated here.

38-4 Locating Events with an Intelligent Observer

In devising special relativity, Einstein stripped science to its bare essentials. The essence of science is the description of *events*—occurrences in space and time. Science has a simple task: to tell us how one event is related to another event. One of the most important outcomes of special relativity is the ability to predict how events observed in one reference frame will look to an observer in another frame. We need to start by carefully defining what events are and how to observe them intelligently.

> An event is an occurrence that happens at a unique place and time.

Examples of events include a collision, an explosion, the emission of a light flash, and the fleeting touch of a friend's hand. When can an occurrence be called an event? When an observer finds it sufficiently localized in space and time to serve her purposes. Your birth was an event unique in both time and place for a genealogist who studies family trees. Your birth mother, however, experienced the process as a *series* of events, from first contraction (maybe at home) to delivery (perhaps in a hospital). Since your birth mother's experiences spanned both time and space, she might not call your birth an event (at least while it is taking place!).

Locating an event in space and time is not always as simple as it might seem at first because of the time delay between event and observation. Think about observing a lightning flash in the night sky. We count the seconds, "one-thousand-one, one-thousand-two, one-thousand-three." Then we hear a crash of thunder. "Wow, lightning struck only one kilometer away from us!" We know this from our knowledge that it takes sound about three seconds to travel one kilometer in air. In making our calculation, we *assume* that the time it takes the lightning flash to reach us is negligible. This means that the lapse between receiving the flash and hearing the thunder is entirely due to the travel time of sound. In this case the signal travels with the speed of sound.

A pulse of high-energy particles may move at nearly the speed of light. How do we determine the time of events that occur along its path? Suppose a pulse of high-energy protons emerges from a particle accelerator and passes through detector A, where we are standing. The pulse continues its flight to arrive at detector B that lies 30 meters away. *When* did the pulse arrive at detector B? We arrange in advance for detector B to send us a light flash when the pulse arrives there. We time the arrival of this light flash at detector A and from this arrival time we *subtract* the known time delay that results when the light flash travels 30 meters. This *difference* gives us the time at which the pulse arrived at detector B.

To account for the delay due to the speed of light, we define the **intelligent observer** to be someone who takes into account the time delays required to locate distant events in space and time. Standing by detector A in the example above, we acted as intelligent observers in determining the time at which the pulse reached detector B.

READING EXERCISE 38-3: The Minute Waltz by Friedrich Chopin takes more than a minute for most pianists to perform. Halfway through playing the Minute Waltz at a recital, will you think of your performance as a single event? Is your performance a single event for those who printed the program for the recital? Looking back ten years later, will you think of it as a single event? ■

READING EXERCISE 38-4: When the pulse of protons passes through detector A (next to us), we start our clock from the time $t = 0$ microseconds. The light flash from detector B arrives back at detector A at a time $t = 0.225$ microsecond (0.225×10^{-6} second) later. (a) At what time did the pulse arrive at detector B? (b) Use the result from part (a) to find the speed at which the proton pulse moved, as a fraction of the speed of light. ■

TOUCHSTONE EXAMPLE 38-2: Simultaneous?

You are an intelligent observer standing next to beacon A, which emits a flash of light every 10 s. 100 km distant from you is a second beacon, beacon B, stationary with respect to you, that also emits a light flash every 10 s. You want to know whether or not each flash is emitted from remote beacon B simultaneous with (at the same time as) the flash from your own beacon A. Explain how to do this without leaving your position next to beacon A. Be specific and use numerical values. Assume that light travels 3×10^8 m/s.

SOLUTION ■ You are an intelligent observer, which means that you know how to take into account the speed of light in determining the time of a remote event, in this case the time of emission of a flash by the distant beacon B. You measure the time lapse between

emission of a flash by your beacon A and your reception of the flash from beacon B. If this time lapse is just that required for light to move from beacon B to beacon A, then the two emissions occur at the same time. The two beacons are 100 km = 10^5 m apart. Call this distance L. Then the time Δt for a light flash to move from B to A is:

$$t = \frac{L}{c} = \frac{10^5 \text{ m}}{3 \times 10^8 \text{ m/s}} = 3.33 \times 10^{-4} \text{ s}, \quad \text{(Answer)}$$

or 0.333 ms. If this is the time you record between the flash of nearby beacon A and reception of the flash from distant beacon B, then you are justified in saying that the two beacons emit their flashes simultaneously in your frame.

38-5 Laboratory and Rocket Latticeworks of Clocks

There are difficulties with the procedure used by our intelligent observer. First, she needs to make a separate calculation for each remote event. This is bothersome. Second, and more fundamental, she cannot calculate the time delay in reporting remote events unless she *already knows the location of every event she wants to measure.* Sometimes information about event location is easily available, sometimes not. We need a general, conceptually simple way to observe both the location and time of events.

In principle, one way to do this is to assemble a cubical lattice of meter sticks with a recording clock at each intersection (Fig. 38-3). Using this latticework, we say that the *position* of an event is that of the recording clock nearest to the event. The *time* of the event is the time recorded on that nearby clock. *Observing an event* then reduces to recording the position of the clock nearest to the event and the time for the event recorded on that clock. Now there is no delay in recording the position and the time of any event that occurs in the lattice.

Synchronizing Latticework Clocks

Before we can actually observe events with our latticework of meter sticks and clocks, we need to set all the clocks in the lattice to read the *same time.* But how can we *synchronize* all the clocks in the latticework? One method would be to carry a traveling clock around the lattice and synchronize each lattice clock with it. This approach is not only time-consuming but incorrect. You will see in the next section that a clock traveling through the lattice runs at a different rate than a resting clock as recorded by clocks in the lattice. In fact, if you set a lattice clock to the time of the traveling clock and then later bring it back after it has traveled to other clocks, you will find that the traveling clock no longer agrees with that lattice clock!

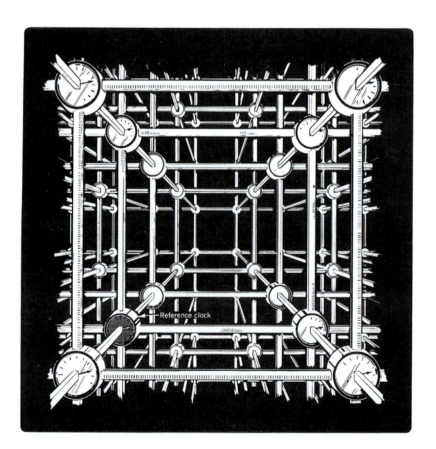

FIGURE 38-3 ■ Latticework of meter sticks and clocks.

Instead of a traveling clock, we use the speed of light to synchronize all the lattice clocks. Our procedure starts by picking one clock in the lattice as the standard or **reference clock.** We know the distance between the reference clock and every other clock in the lattice. At midnight the reference clock sends out a **synchronizing flash** of light. When an observer at any one of the distant clocks receives the flash, she quickly sets the time on her clock to midnight *plus* the time it took for the light to reach her over the known distance from the reference clock at the known speed of light. We say that after this procedure is complete the clocks in the lattice read the *same time* as one another—they are *synchronized* with respect to this lattice.

Now our latticework of synchronized clocks is ready to record the position and time of events that occur during any experiment. Analyzing the results of that experiment means relating events by collecting event data from all recording clocks in the lattice and analyzing these data at some central location.

Laboratory and Rocket Frames as Inertial Reference Frames

We often hear talk in special relativity about the *laboratory frame* and the *rocket frame.* Envision *each* of these frames as having a latticework of rods and clocks (Fig. 38-4). The rocket coasts at constant velocity in unpowered flight. By convention we assign the positive *x* direction to be the direction of motion of the rocket with respect to the laboratory lattice.

> **AN IMPORTANT ASIDE:** Strictly speaking, reference frames used in special relativity must be inertial frames, frames with respect to which Newton's First Law of motion holds: *A free particle at rest remains at rest and a free particle in motion continues that motion in a straight line at constant speed* (see Section 3-2). Obviously the surface of the Earth is not an inertial reference frame; a stone released from rest accelerates downward! However, for a particle moving at a substantial fraction of the speed of light with respect to the Earth, the acceleration of gravity can usually be ignored. In this chapter we make no distinction between inertial frames and those at rest or moving at constant velocity with respect to the Earth's surface.

We can detect and record a single event using overlapping rocket and laboratory lattice works. If the right rear tire of your car hits a nail, it goes flat with a *bang.* For you as the driver (in the "rocket frame") the bang occurs at the right rear of your car. For the observer on the road (the "laboratory frame"), the bang takes place where the nail sticks up at one end of the bridge that your car has just crossed. Neither you nor the road observer "owns" the event. You both have equal status in observing and recording the bang. The bang exists, *and all other events exist,* independent of reference frames. Events are the nails on which all of science hangs.

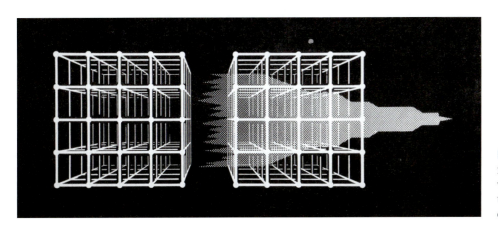

FIGURE 38-4 ■ Laboratory and rocket frames. A moment ago the two latticeworks were intermeshed. By convention, the rocket frame moves in the positive *x* direction of the laboratory frame.

You are stationed at a latticework clock with the coordinates $x = 3 \times 10^8$ m, $y = 4 \times 10^8$ m, and $z = 0$ m. The reference clock at coordinates $x = y = z = 0$ emits a reference flash at exactly midnight on its clock. You want your clock to be synchronized with (set to the same time as) the reference clock. To what time do you immediately set your clock when you receive the reference flash?

SOLUTION ■ Your distance D from the reference clock is

$$D = [(3 \times 10^8 \text{ m})^2 + (4 \times 10^8 \text{ m})^2 + 0 \text{ m}]^{1/2} = [25 \text{ m}]^{1/2} \times 10^8$$

$$= 5 \times 10^8 \text{ m}.$$

The time Δt that it takes the reference flash to reach you is therefore

$$\Delta t = \frac{D}{c} = \frac{5 \times 10^8 \text{ m}}{3 \times 10^8 \text{ m/s}} = 1.66 \text{ s}. \qquad \text{(Answer)}$$

So when you receive the reference flash, you quickly set your clock to 1.66 seconds after midnight.

38-6 Time Stretching

Every year hundreds of email messages, letters, and articles "disproving" relativity are sent to textbook authors and scientific journal editors. Many of these papers are extremely ingenious, showing considerable insight and sometimes representing years of labor. (Indeed, fighting a new idea often helps us to understand it and make it our own. As you continue reading this chapter, you may want to make a note of the ideas that seem paradoxical or outrageous. By the time you finish the chapter, see if you can refute or defend some of your initial objections to relativity.)

A primary target of writers who object to special relativity is **time stretching,** the conclusion that the time between two events can have different values as measured in laboratory and rocket frames in relative motion. The clearest case of time stretching is this: *The time between two ticks measured on a clock at rest is always less than the time between the same two ticks measured in a reference frame in which the clock is moving.* Many people remember this result by using a not-quite-exact motto: *Moving clocks run slow.* However we express it, this conclusion is so obviously ridiculous that it stimulates dozens of skeptics to write letters and articles.

Verification of Time Stretching

Time stretching is verified experimentally every day as part of the ongoing enterprise of experimental physics. Here are two examples of time stretching in action.

Time stretching with atomic clocks: In October 1971, J. C. Hafele and R. E. Keating of the U.S. Naval Observatory sent atomic clocks (like the one described in Section 1-5) around the Earth on regularly scheduled commercial airliners. One clock circled the globe traveling eastward, the other clock traveled westward. When the clocks were finally brought together, they did not read the same time. Also, the reading on both clocks was different from that of a third atomic clock, which stayed at home in one place on the Earth's surface. Why the different readings? Think of the center of the Earth as at rest. (Actually, the center of the Earth is in free fall around the sun.) With respect to the Earth's center, the speed of the eastward-moving clock is *added* to the speed of the Earth's rotation; it is the "faster-moving" clock. In contrast, the speed of the westward-moving clock is *subtracted* from the eastward motion of the Earth's surface; this is the "slower-moving" clock. The stay-at-home clock moves with the Earth's surface at a speed that is intermediate between that of the other two clocks. The result? The "faster-moving" eastward-going clock runs slow compared with the stay-at-home clock of intermediate speed. And the "slower-moving" westward-going clock runs fast compared with the stay-at-home clock. In the Hafele–Keating experiment, the magnitudes of the different readings corresponded to the predictions of relativity. There was, however, at least one complication: The airplanes changed altitude as they took off, flew their courses, and landed. General relativity, the theory that includes gravitational effects, predicts that changes in altitude, as well as relative speeds, affect the relative rates at which clocks run. The results of the Hafele–Keating experiment were actually consistent with the predictions of general relativity too, but that is a story for another day.

Time stretching with pions: Our second example of time stretching is more technical than the Hafele–Keating experiment, but a lot more convincing. It involves measuring the lifetimes of **pions,** also called pi-mesons or π^+ mesons. These short-lived particles can be created during cosmic ray interactions or when a beam of protons energized by a particle accelerator strikes a target. On average, half the pions in a beam will decay into other particles in 18 nanoseconds (18×10^{-9} seconds) as measured by a clock carried with the pions. In this pion frame, half of the remainder will decay in the next 18 nanoseconds, and so on. We call this time the pion **half-life** ($t_{1/2}$). If pions are moving at nearly the speed of light, how far can a pion beam travel before half the pions decay? If the time were the same in our laboratory as it is in the rest frame of the speeding pions, the maximum distance would be approximately equal to

$$c \times t_{1/2} = (3 \times 10^8 \, \text{m/s}) \times (18 \times 10^{-9} \, \text{s}) = 5.4 \, \text{m}.$$

However, experiment shows that the flying pions travel tens of meters before half of them decay. We conclude that in our laboratory frame the time for half of the pions to decay is much greater than it is in the rest frame of the pions. Time stretching!

Why Time Stretching Makes Sense

Objections to time stretching have always failed because they attack a result based on an utterly simple idea: All the laws of nature are the same in every reference frame (the Principle of Relativity). In particular, the speed of light is invariant (that is, it has the same value) in every reference frame. The invariance of the speed of light leads directly to the difference in time between two events as measured in laboratory and rocket frames. To illustrate this, let's consider the ticking of a "light clock" diagrammed in Fig. 38-5.

While riding in a transparent unpowered rocket ship, you fire a flash of light upward toward a mirror that you hold on a stick 3 meters directly above you (the left-hand panel in Fig. 38-5). The flash reflects at the mirror and returns to you. Call the emission of light event A and its reception upon return event B. For you, events A and B occur at the same place. Between the events, the light moves first straight up

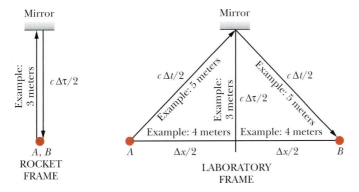

FIGURE 38-5 ▪ A flash of light emitted at event A reflects from a mirror and returns to the source, arriving as event B. Events A and B are recorded in both the rocket and laboratory frames. Einstein tells us that each observer measures the same speed of light. Therefore different path lengths for the light flash in rocket and laboratory frames mean that different times between events A and B are measured in the two frames. (The meanings of the symbols Δx, $c \, \Delta t$, and $c \, \Delta \tau$ are discussed in the text.)

3 meters then straight down 3 meters. *For you the total time between events A and B equals the time that it takes light to travel a total of 6 meters.*

The unpowered spaceship in which you carry out these experiments moves from left to right past the rest of us, who stand in another transparent container arbitrarily labeled "laboratory" (the right-hand panel in Fig. 38-5). We also observe the same flash of light emitted at event *A*, reflected at your mirror, and received again at event *B*. But for us in the laboratory, you and your mirror move together to the right. Thus for us the path of the light slants upward from *A* along the 5-meter-long (for example) hypotenuse of the first right triangle, reflects from the speeding mirror, then slants back down along the 5-meter hypotenuse of the second right triangle to meet you again at event *B*. Therefore *for us events A and B are separated by the time it takes light to travel a total of 10 meters.* For us a longer time lapses between events *A* and *B* than you measured in your rocket frame.

That's it! Longer path length for light, longer time for light to travel that path at its "universal speed," therefore longer time between events as measured in that reference frame. No one has ever found an acceptable way around this simple and powerful result. The light clock demonstrates the longer time between two events in one frame than in another frame. Hence the name for this effect is **time stretching** or **time dilation** (*dilation* is a medical term for *stretching*).

The light flash and mirror make a kind of clock that we define as a **light clock.** The Principle of Relativity assures us that *all kinds of clocks* at rest in a frame, once calibrated, must run at the same rate as one another as observed in every frame in uniform relative motion with respect to the first frame. Otherwise we could tell which frame we are in by detecting different rates of different clocks all at rest in our frame. In any given frame, *properly calibrated clocks of every kind run at the same rate as one another,* including the "clock" of your body—namely, the aging process. Suppose the mirror was so high above your head that it took 6 years in your rocket for the light to return to you—and 10 years by our laboratory clocks. Then you would age 6 years between these new events *A* and *B* because your body's "aging clock" and your "light clock" ride together in your rocket frame. In contrast, between events *A* and *B* we in the laboratory frame would age 10 years, and our light clock would also advance 10 years. Between events *A* and *B* you would age less than we do!

How strange it is that the speed of light is *invariant* for all observers, no matter what their relative velocity! But experiment continually verifies this result. Consequences of the invariance of the speed of light include the fact that clocks run at different rates for observers in relative motion. Experiment continually verifies this result as well. More than one hundred years of the most rigorous testing have validated beyond reasonable doubt that the speed of light is the same in all reference frames and that clocks tick at different rates when observed in frames that are in motion with respect to each other.

READING EXERCISE 38-5: Suppose that a beam of pions moves so fast that at 25 meters from the target in the laboratory frame exactly half of the original number remain undecayed. As an experimenter, you want to put more distance between the target and your detectors. You are satisfied to have one-eighth of the initial number of pions remaining when they reach your detectors. How far can you place your detectors from the target? ∎

READING EXERCISE 38-6: A set of clocks is assembled in a stationary boxcar. They include a quartz wristwatch, a balance wheel alarm clock, a pendulum grandfather clock, a cesium atomic clock, fruit flies with average individual lifetimes of 2.3 days, a clock based on radioactive decay of nuclei, and a clock timed by marbles rolling down a track. The clocks are adjusted to run at the same rate as one another. The boxcar is then gently accelerated along a smooth horizontal track to a final velocity of 300 km/hr. At this constant final speed, which clocks will run at a different rate from the others as measured in that moving boxcar? ∎

38-7 The Metric Equation

Time stretching occurs whenever the rocket has a velocity with respect to the laboratory frame. We can prove this in general using symbols in Fig. 38-5. In the laboratory frame we measure Δx as the distance between events A and B and we measure Δt as the time between the events (that is, the time it takes the light flash to slant upward along one hypotenuse at the speed c and then slant downward to the second event). Here delta (Δ) indicates the *lapse* of time and the letter t tells us that the elapsed time refers to the time between two events in our laboratory frame. One hypotenuse has a length given by

$$\text{length} = \text{velocity} \times \text{time} = c\,\Delta t/2.$$

In your rocket frame the flash moves vertically upward to the mirror and back down again in the time between events A and B. To describe your rocket frame time lapse between events we use the notation $\Delta\tau$. Here delta (Δ) indicates the *lapse* of time and the Greek letter tau (τ) tells us that the elapsed time refers to the time between two events in your unpowered rocket, in which they occur at the same place. So for you the upward distance covered in time $\Delta\tau/2$ is equal to $c\,\Delta\tau/2$. This vertical span is the same as the vertical leg shared by the right triangles in the diagram at the right. Hence we have expressions for the lengths of all sides of both of these right triangles. If we use the Pythagorean Theorem for right triangles we get

$$(\Delta x)^2 + (c\,\Delta\tau)^2 = (c\,\Delta t)^2 \tag{38-1}$$

where each term has dimensions of length squared. If we rearrange the terms in Eq. 38-1 so all the terms that refer to the laboratory frame are on the right, we get a squared time-like interval called the **metric equation:**

$$(c\,\Delta\tau)^2 = (c\,\Delta t)^2 - (\Delta x)^2 \qquad \text{(squared time-like interval).} \tag{38-2}$$

Now suppose two events occur in the same place ($\Delta x = 0$), such as two sequential ticks of a clock in its rest frame. The time lapse $\Delta\tau$ between the events measured on the clock at rest is called the **proper time** or **wristwatch time.** The German term for proper time is *Eigenzeit,* meaning "one's own time." The square root of the difference of squares on the right side of Eq. 38-2 has the formal name **invariant time-like interval.** The interval is *invariant* because it has the same value as calculated by all observers. It is *time-like* because the magnitude of the time part $c\,\Delta t$ is greater than the magnitude of the space part Δx. (Both sides of Eq. 38-2 are necessarily positive).

The metric equation 38-2 is one of the most amazing equations in all of physics. Look at its outrageous implications:

- First, the metric equation relates *two different* measures of the time between the *same* two events. These are: (1) the time recorded on clocks in the reference frame in which the events occur at different places, and also (2) the wristwatch time read on the clock carried by a traveler who records the two events as occurring at the same place. The ability to relate these two times is one of the greatest scientific innovations in history.

- Second, the metric equation reveals an even deeper insight—space and time combine in a single expression on the right side. We no longer speak of space and time separately, but as a unity: **space-time**!

A wealth of other insights can be gleaned from Eq. 38-2. For example:

1. The time between events A and B as measured in the two frames *cannot have the same value* if the laboratory and rocket frames are in relative motion.

2. Laboratory observers can correctly predict the proper time observers in the rocket frame measure between events A and B on their rocket-frame wristwatches, in spite of the fact that this time is not the same as the time measured in the laboratory frame. (Observers in the laboratory simply put their values for Δt and Δx into Eq. 38-2 and calculate the value of the wristwatch time $\Delta\tau$ that the rocket observer measures.)

3. If the rocket speed relative to the laboratory frame is reduced, then both $\Delta x/2$ (the length of the horizontal leg in Fig. 38-5) and $c\,\Delta t/2$ (the hypotenuse) will be smaller than before. But these terms become smaller in such a way that the difference between their squares, which represents the vertical distance between the rocket observer and her mirror $c\,\Delta\tau/2$, will remain the same. So the metric equation (Eq. 38-2) will still hold. No matter how fast or slow the rocket is, the value of the proper time $\Delta\tau$ (also known as the invariant time-like interval) remains the same. Hence we call this interval between two events **invariant**, meaning that it has the same value as measured in *all* reference frames in uniform relative motion.

If a rocket passes by our laboratory frame at a speed v, we can derive an equation that relates $\Delta\tau$ and Δt directly by setting $\Delta x = v\,\Delta t$. Substituting this expression into Eq. 38-2 and dividing through by c^2 gives us

$$(\Delta\tau)^2 = (\Delta t)^2 - \left(\frac{\Delta x}{c}\right)^2 = (\Delta t)^2 - \left(\frac{v\,\Delta t}{c}\right)^2 = \left(1 - \frac{v^2}{c^2}\right)(\Delta t)^2.$$

Taking the square root of both sides gives us an expression known as the *time-stretching* or *time dilation equation,*

$$\Delta\tau = \sqrt{1 - v^2/c^2}\,\Delta t \qquad \text{(time-stretching equation).} \qquad (38\text{-}3)$$

The time-stretching equation (Eq. 38-3) gives us the value of wristwatch time $\Delta\tau$ between two events that occur a time Δt apart in some reference frame. In this equation, v is the speed required for an observer in the rocket frame to move directly from one event to the other event. The equation encompasses all possible values of speed v from the very slow to the very fast.

What Happens at High and Low Speeds?

The speeds we observe in everyday life are so much smaller than the speed of light c that the value of v/c is extremely small compared to 1. Thus, for low relative speeds, the expression $(1 - v^2/c^2)$ is approximately equal to 1. The time-stretching equation (Eq. 38-3) tells us that in this case Δt and $\Delta\tau$ are essentially equal; the time between events A and B is the same for you in a passing airplane as it is for us standing on Earth. So at very low relative speeds special relativity is consistent with our everyday assumption that time is a universal quantity, that everyone measures the time between two events to have the same value. This is the approximating assumption used in Newtonian mechanics.

In contrast, at a high relative speed v the outcome is quite different from what happens at everyday speeds. Imagine that you start from Earth (event A: departure from Earth) and travel to the star Alpha Centauri, about 4 light-years from Earth (event B: arrival at Alpha Centauri). Both events A and B (departure and arrival) occur at the position of your cockpit. Equation 38-3 tells us that by making v/c closer and closer to the value unity, your trip can take place in shorter and shorter wristwatch time $\Delta\tau$ *as measured in your spaceship.* (This is true even though the time Δt measured in the Earth frame can never be less than the time it takes light to move from Earth to Alpha Centauri—4 years.) By extension of this argument, we arrive at a result that frees the human spirit, if not yet the human body. Given sufficient rocket speed, we can go anywhere in the universe in the lifetime of a single astronaut! At least this is the prediction of special relativity.

Nature's Speed Limit

The time-stretching equation (Eq. 38-3) also gives evidence that the natural speed limit of the universe is the speed of light c. Imagine that we in the laboratory measured your rocket speed v to be greater than the speed of light c. Then v/c (and also v^2/c^2) would have a value greater than unity, and the expression on the right side of Eq. 38-3 would include the square root of a negative number. This would mean that the time measurement $\Delta\tau$ would be proportional to the square root of a negative number. But this is impossible: No real time can be proportional to the square root of a negative number. Careful study and experiment have led to the conclusion, consistent with this formula, that no object can be accelerated to a speed v greater than the speed of light c in a vacuum. Experiment verifies this: Many nations together have spent billions of dollars to build and operate huge particle accelerators that use electric and magnetic fields to urge protons or electrons to ever-higher energies. At higher and higher energies, these particles approach closer and closer to the speed of light but have never been observed to exceed this speed.

READING EXERCISE 38-7: Find the rocket speed v at which the time $\Delta\tau$ between ticks on the rocket clock is recorded by the laboratory clock as $\Delta t = 1.01\,\Delta\tau$. ◼

TOUCHSTONE EXAMPLE 38-4: Satellite Clock Runs Slow?

An Earth satellite in circular orbit just above the atmosphere circles the Earth once every $T = 90$ min. Take the radius of this orbit to be $r = 6500$ kilometers from the center of the Earth. How long a time will elapse before the reading on the satellite clock and the reading on a clock on the Earth's surface differ by one microsecond? For purposes of this approximate analysis, assume that the Earth does not rotate and ignore gravitational effects due the difference in altitude between the two clocks (gravitational effects described by general relativity).

SOLUTION ◼ First we need to know the speed of the satellite in orbit. From the radius of the orbit we compute the circumference and divide by the time needed to cover that circumference:

$$v = \frac{2\pi r}{T} = \frac{2\pi \times 6500 \text{ km}}{90 \times 60 \text{ s}} = 7.56 \text{ km/s}. \quad (38\text{-}4)$$

Light speed is almost exactly $c = 3 \times 10^5$ km/s, so the satellite moves at the fraction of the speed of light given by

$$\frac{v}{c} = \frac{7.56 \text{ km/s}}{3 \times 10^5 \text{ km/s}} = 2.52 \times 10^{-5}. \quad (38\text{-}5)$$

or

$$v^2/c^2 = (2.52 \times 10^{-5})^2 = 6.35 \times 10^{-10}. \quad (38\text{-}6)$$

The relation between the time lapse $\Delta\tau$ recorded on the satellite clock and the time lapse Δt on the clock on Earth (ignoring the Earth's rotation and gravitational effects) is given by Eq. 38-3. Square both sides of that equation to obtain:

$$(\Delta\tau)^2 = (1 - v^2/c^2)(\Delta t)^2. \quad (38\text{-}7)$$

We want to know the difference between Δt and $\Delta\tau$. Rearrange this equation to give the difference of squares:

$$v^2/c^2(\Delta t)^2 = (\Delta t)^2 - (\Delta\tau)^2 \equiv (\Delta t - \Delta\tau)(\Delta t + \Delta\tau). \quad (38\text{-}8)$$

Substituting the numerical result of Eq. 38-6 into Eq. 38-7, we see that $\Delta\tau$ and Δt have very nearly the same value. Therefore we can set

$$\Delta t + \Delta\tau \approx 2\Delta t. \quad (38\text{-}9)$$

With this substitution, Eq. 38-8 becomes

$$v^2/c^2(\Delta t/2) \approx \Delta t - \Delta\tau. \quad (38\text{-}10)$$

Substitute from Eq. 38-6:

$$\Delta t - \Delta\tau \approx 3.18 \times 10^{-10}\Delta t. \quad (38\text{-}11)$$

We are asked to find the elapsed Δt for which the satellite clock and the Earth clock differ in their reading by one microsecond = 10^{-6} second. Rearrange Eq. 38-11 to read

$$\Delta t \approx \frac{\Delta t - \Delta\tau}{3.18 \times 10^{-10}} = \frac{10^{-6} \text{ s}}{3.18 \times 10^{-10}} = 3.14 \times 10^3 \text{ s}. \quad \text{(Answer)}$$

This is approximately equal to 52 minutes, or a little less than one hour. A difference of one microsecond between atomic clocks is easily detectable.

38-8 Cause and Effect

The analysis thus far has omitted from our consideration a large number of possible pairs of events. Suppose two events occur at the same time but not at the same place in a reference frame. For example, what if two firecracker explosions occur simultaneously, one in New York City, the other in San Francisco? Since $\Delta\tau = 0$ for this pair of events, Eq. 38-2 for the space-time interval becomes

$$(c\,\Delta\tau)^2 = (c\,\Delta t)^2 - (\Delta x)^2 \rightarrow -(\Delta x)^2 \qquad \text{(for } \Delta t = 0\text{)}.$$

What can this expression possibly mean? The left side contains the square of a time, obviously a positive quantity. Yet on the right is a negative quantity. No clock records a time lapse $\Delta\tau$ whose square is a negative quantity! We have a contradiction here, and a contradiction that applies in a similar way to all possible pairs of events simultaneous in some frame.

The problem is not with physics but with mathematical notation. Pairs of simultaneous events were not envisioned in the derivation of the proper time equation (Eq. 38-2) based on Fig. 38-5. For this new class of event-pairs we need a new formalism. To achieve this, reverse the order of squared quantities on the right side of Eq. 38-2 and give the result a different name. Earlier we used the notation $\Delta\tau$ (involving the Greek letter tau—denoted τ) to represent the elapsed *time* between two events measured on a clock for which the events occur at the same *place*. For our new expression we use the notation $\Delta\sigma$ (involving the Greek letter sigma—denoted σ) to represent the *distance* between two events measured in the frame in which they occur at the same time. This new equation has the form

$$(\Delta\sigma)^2 = (\Delta x)^2 - (c\,\Delta t)^2 \qquad \text{(squared space-like interval)}. \qquad (38\text{-}12)$$

The distance $\Delta\sigma$ between two events, measured in a frame in which the events occur at the same time ($\Delta t = 0$), is called the **proper distance.** This square root of the difference of squares on the right side of Eq. 38-12 also has the formal name **invariant space-like interval**—space-like because the space part Δx is greater than the time part $c\,\Delta t$.

The right-hand side of Eq. 38-12 also describes the space and time separations between these two events as measured in a second frame that moves past the first; in the second frame Δx and Δt are both different from zero. The proper distance $\Delta\sigma$, like the wristwatch time $\Delta\tau$, is an *invariant* in the following sense: Observers in relative motion may measure different values of Δx and different values of Δt between these two events. However, when each observer substitutes these values into Eq. 38-12, he will obtain the same numerical value for the proper distance $\Delta\sigma$. And the value of $\Delta\sigma$ is just the distance between the two events as measured in that particular reference frame in which they occur at the same time.

Some important consequences for events separated by a space-like interval can be read from Eq. 38-12:

- If Δx is greater in magnitude than $c\,\Delta t$ in one frame, then Δx is greater in magnitude than $c\,\Delta t$ in all frames. Why? Because $\Delta\sigma$ is an invariant, so $(\Delta\sigma)^2$ has the same value whatever the values of Δx and $c\,\Delta t$ in a particular frame. Both sides of Eq. 38-12 must remain positive. But $c\,\Delta t$ is the distance that light can travel in the time available between these events. Equation 38-12 says that Δx is greater than this distance $c\,\Delta t$ between the two events in that frame. Nothing, not even a light flash, can move fast enough to travel from one event to the other in the elapsed time Δt between them. Therefore, for events connected by a space-like interval, one event *cannot* cause the other event as observed in *any* frame.

- By definition, $\Delta\sigma$ is the separation between two events in a reference frame in which the time between these events is zero so the events are simultaneous. But the right side of Eq. 38-12 contains *both* Δt *and* Δx, implying that for another frame in relative motion Δt is *not* zero. That is, in this other frame the two events are *not* simultaneous. As an example, for observers in a rocket streaking across the continent from New York toward San Francisco, the two firecracker explosions in New York and San Francisco will *not* occur at the same time. This leads to a major result of special relativity: *Two events simultaneous in one frame are not necessarily simultaneous in other frames in motion relative to the first.* For many people this is the most difficult concept in special relativity, harder to believe even than the difference in clock rates described by the time-stretching equation (Eq. 38-3). In Section 38-9 we elaborate on this result, which is called the **relativity of simultaneity.**

Suppose that two events are separated in space Δx and time Δt so that a flash of light moving directly between them can *just* make it from one event to the other event in the time Δt. Then the distance between them is given by $\Delta x = c\,\Delta t$. In this case both the proper distance $\Delta\sigma$ and the proper time $\Delta\tau$ are zero:

$$(c\,\Delta\tau)^2 = (\Delta\sigma)^2 = 0 \qquad \text{(squared light-like interval).} \qquad (38\text{-}13)$$

Two events that can be connected by a direct light flash are said to be related by an **invariant light-like interval** or **null interval**—null because the space part Δx is equal in magnitude to the time part $c\,\Delta t$, so the difference between their squares is zero, or null.

   Equations 38-2, 38-12, and 38-13 embrace all possible cause-and-effect relations between pairs of events that occur along the x direction as described by special relativity. Equation 38-2 describes two events separated by a time-like interval. Something moving more slowly than light, a rocket for example, can travel directly between these two events in the time between them, so it is possible for the earlier event to cause the later event. This possible cause-and-effect relation between an earlier and a later event is preserved in every reference frame. In contrast, not even light can travel between the two events separated by a space-like interval described in Eq. 38-12, so that neither one of these two events can cause the other event. This *lack* of possible cause-and-effect relation is preserved in every frame. Equation 38-13 provides the boundary between these two cases: the relation between two events that can be connected only by a direct light flash. The earlier event in this pair can cause the other event only through a directly connecting light flash. This cause-and-effect relation between the earlier and later events is also maintained in every reference frame.

   In brief, the three-fold categories of time-like, space-like, and light-like intervals between pairs of events preserve the possible cause-and-effect relation between these events in *all* reference frames. Special relativity may be weird, but at least it reaffirms the fact that cause comes before effect for all observers—a statement that most of us consider to be a central requirement of science.

READING EXERCISE 38-8: Points on the surfaces of the Earth and the Moon that face each other are separated by a distance of 3.76×10^8 meters. How long does it take light to travel between these points? A firecracker explodes at each of these two points; the time between these explosions is one second. Is it possible that one of these explosions caused the other explosion? ■

38-9 Relativity of Simultaneity

In the previous section we obtained the following result from the space-like form of the metric equation:

Two events that are simultaneous in one frame are not necessarily simultaneous in a second frame in uniform relative motion.

This result becomes clear when we consider what observers on the ground and the train see as each measures the time between the same two events, as shown in Fig. 38-6. Suppose lightning bolts strike both ends of the train, emitting flashes and leaving char marks on both the train and the track (top image in the figure). Assume that flashes from the front and back of the train reach the observer on the ground at the same time (bottom image in the figure). This ground observer measures his distance from the two char marks on the track and finds these distances to be equal. He concludes that, for him, the two lightning bolts struck *simultaneously*. In contrast, the rider at the middle of the train sees the flash from the front of the train first (because in Fig. 38-6 she moves toward the light flash coming from the front of the train and away from the light flash coming from the back). She measures her distance from the char marks on the two ends of the train and finds these distances equal. Following the Principle of Relativity, she assumes that the speed of light has the same value in her train frame as in every other frame. She concludes that, for her, the lightning struck the front end of the train first. Her reasoning is explained in the caption to Fig. 38-6.

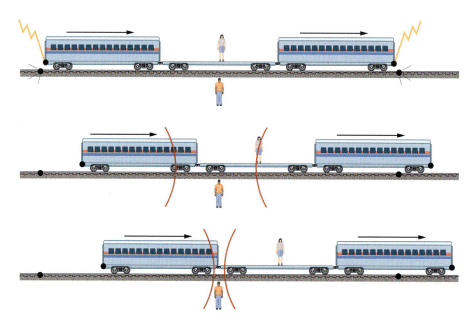

FIGURE 38-6 ■ Einstein's Train Paradox illustrating the relativity of simultaneity. *Top:* Lightning strikes the front and back ends of a moving train, leaving char marks on both track and train. Each emitted flash spreads out in all directions. *Center:* Observer riding in the middle of the train concludes that the two strokes are *not* simultaneous. Her argument: "(1) I am equidistant from the front and back char marks on the train. (2) Light has the standard speed in my frame, and equal speed in both directions. (3) The flash arrived from the front of the train first. (4) Therefore, the flash must have left the front of the train first; the front lightning bolt fell before the rear lightning bolt fell. I conclude that the lightning strokes were not simultaneous." *Bottom:* Observer standing by the tracks halfway between the char marks on the tracks concludes that the two lightning strokes were simultaneous, since the flashes from the strokes reach him at the same time and he is equidistant from the char marks on the track. *Conclusion:* Two events that are simultaneous in one frame may *not* be simultaneous in another frame.

READING EXERCISE 38-9: Susan, the rider on the train pictured in Fig. 38-6, is carrying an audio tape player. When she receives the light flash from the front of the train she switches on the tape player, which plays *very loud* music. When she receives the light flash from the back end of the train, Susan switches off the tape player. Will Sam, the observer on the ground, be able to hear this music? Later Susan and Sam meet for coffee and examine the tape player. Will they agree that some tape has been wound from one spool to the other? ■

TOUCHSTONE EXAMPLE 38-5: Principle of Relativity Applied

Divide the following items into two lists. On one list, labeled SAME, place items that name properties and laws that are always the *same* in every frame. On the second list, labeled MAY BE DIFFERENT, place items that name properties that can be *different* in different frames:

a. the time between two given events
b. the distance between two given events
c. the numerical value of Planck's constant h
d. the numerical value of the speed of light c
e. the numerical value of the charge e on the electron
f. the mass m of an electron (measured at rest)
g. the elapsed time on the wristwatch of a person moving between two given events
h. the order of elements in the periodic table
i. Newton's First Law of Motion ("A particle initially at rest remains at rest, and . . .")
j. Maxwell's equations that describe electromagnetic fields in a vacuum
k. the distance between two simultaneous events

SOLUTION ■ The Principle of Relativity says that the laws of physics are the same in every frame. So items (i) and (j) should go on the SAME list, along with item (h). The Principle of Relativity extends to the values of fundamental constants, so items (c), (d), (e), and (f) should also go on the SAME list.

In contrast, as we have seen in this chapter, the time between a pair of events (item a) may be different in different frames. The same is true for the distance between two events (item b). So these go in the DIFFERENT list.

This leaves two items, (g) and (k). Item (g), the time on the wristwatch of a person moving between two given events (the so-called "wristwatch time") is an invariant, the same as calculated using space and time separations measured in any frame (Eq. 38-2). So this goes on the SAME list. The same is true of item (k), the "proper distance" between two events. This is also an invariant and goes on the SAME list.

In summary, here are the two lists requested: (Answer)

THE SAME IN ALL FRAMES	MAY BE DIFFERENT IN DIFFERENT FRAMES
c. numerical value of h	a. time between two given events
d. numerical value of c	b. distance between two given events
e. numerical value of e	
f. mass of electron (at rest)	
g. wristwatch time between two events	
h. order of elements in the periodic table	
i. Newton's First Law of Motion	
j. Maxwell's equations	
k. distance between two simultaneous events	

38-10 Momentum and Energy

Shortly after his first paper on special relativity was published, Einstein submitted a paper that added the most famous equation of all time, $E = mc^2$, to his theory. This equation tells us that every particle in the universe with mass is a storehouse of energy, useful to us provided we can find ways to transform this mass into other forms of energy. The explosion of a nuclear weapon and burning of a star provide spectacular examples of transformations of mass to energy, but every single energy-emitting reaction—down to the burning of a match—carries with it a conversion of mass to a significant amount of energy. For example, the wood in a kitchen match contains about 30,000 calories (or 30 food calories). Because of the huge magnitude of the conversion factor c^2, the corresponding predicted change in mass of the combustion products is less than 2 billionths of a gram.

Where Does $E = mc^2$ Come From?

How does the famous $E = mc^2$ equation grow out of the special theory of relativity discussed so far in this chapter? The connection is not direct. In this section we shall present arguments for the development of $E = mc^2$ using equations we have already introduced in this and earlier chapters. We shall also explore some of the consequences of the equivalence of mass and energy. Please be patient and follow the logic. It will be rewarding.

Imagine that a moving particle emits two flashes a time $\Delta\tau$ apart as recorded on its own wristwatch. We use these two emissions to track the motion of the particle. These two flashes can be related using the metric equation (38-2):

$$(c\,\Delta\tau)^2 = (c\,\Delta t)^2 - (\Delta x)^2$$

where the values of Δx and Δt are measured with respect to the laboratory frame. Starting with this equation, we can extract some important information about the momentum and energy of the particle. We start by multiplying both sides of the equation by $m^2 c^2/(\Delta\tau)^2$, where m is the mass of the particle. This gives us

$$(mc^2)^2 = \left(mc^2\,\frac{\Delta t}{\Delta\tau}\right)^2 - \left(mc\,\frac{\Delta x}{\Delta\tau}\right)^2. \tag{38-14}$$

Note that the famous expression mc^2 appears on the left. The second term on the right contains the fraction $\Delta x/\Delta\tau$—namely, the distance Δx traveled by the particle as measured by our laboratory observer, divided by the time $\Delta\tau$ it takes to move this distance as recorded on the wristwatch carried by the particle. This measures a kind of velocity. Mass times velocity yields the formula for momentum; call it p. The laboratory observer reckons the momentum to have the value

$$p \equiv m\,\frac{\Delta x}{\Delta\tau} \qquad \text{(lab observer's definition of particle momentum).} \tag{38-15}$$

But, why does the lab observer use $\Delta\tau$ in Eq. 38-15 rather than Δt to define momentum? Newtonian mechanics assumes that the time Δt between two events is a universal quantity, with the same value as measured in all reference frames. But relativity shows us (Eq. 38-3) that the time between the two flashes emitted by the particle has a different value when measured in different frames. We have chosen to use the invariant proper time $\Delta\tau$ (as recorded on the wristwatch carried by the particle) to be the time to use in reckoning the particle's momentum. So Eq. 38-15 results from a decision about time that Newton did not have to make.

What about the first term on the right side of Eq. 38-14, the one containing the squared ratio of time lapses $(\Delta t/\Delta\tau)^2$? According to Eq. 38-3, this squared ratio is related to the ratio of the particle velocity and the speed of light by the equation

$$\left(\frac{\Delta t}{\Delta\tau}\right)^2 = \frac{1}{1 - v^2/c^2}. \tag{38-16}$$

The $(mc^2)^2$ term on the left in Eq. 38-14 has the units of energy squared. Some powerful results follow if we assume that the first term on the right of that equation is the square of the total energy, E, of the particle. Then, using Eqs. 38-14 and 38-16, energy can be written in two ways:

$$E = mc^2\frac{\Delta t}{\Delta\tau} = \frac{mc^2}{(1 - v^2/c^2)^{1/2}}. \tag{38-17}$$

We can substitute the definition of the momentum, p, from Eq. 38-15 and the definition of energy, E, from Eq. 38-17 into Eq. 38-14 to get

$$(mc^2)^2 = E^2 - (pc)^2. \tag{38-18}$$

When the particle is at rest, the momentum $p = 0$ and this equation takes the famous form

$$E_{rest} = mc^2 \quad \text{(rest energy of a particle of mass } m\text{)}. \quad (38\text{-}19)$$

Note that Eq. 38-19 describes only a particle that is *at rest* in a given frame. For a particle in motion, observers in that frame must use Eq. 38-17 to predict its energy.

    Is there experimental evidence that the expressions for energy and momentum derived above have a useful reality? Yes, overwhelming evidence. In analyzing decades of experiments with high-speed particles, conservation of energy and momentum continue to be valid in special relativity *provided* that one uses the *relativistic* expressions for energy and momentum. In analyzing high-speed particle collisions in an isolated system, one adds up the total energy of particles before a collision, using Eq. 38-17 for each particle (being sure to include the rest energy of any particles at rest). This number will be equal to the total energy of the system of particles after the collision, no matter how many particles are destroyed or created in the process. A similar conservation law holds true for *each* spatial component of the total momentum of the particles, using Eq. 38-15 for the *x*-component and similar equations for other directions (substituting Δy or Δz for Δx).

Relativistic Kinetic Energy

When a given particle is *not* at rest in a given reference frame, then its momentum p is *not* zero as measured in that frame. In this case the total particle energy, E, must be greater than its rest value to keep the right side of Eq. 38-18 a constant, equal to the left side. The *increase in energy of a particle due to its motion* is called **kinetic energy.** In special relativity the kinetic energy is defined as the difference between the total energy and the rest energy:

$$K \equiv E - E_{rest} = \frac{mc^2}{(1 - v^2/c^2)^{1/2}} - mc^2. \quad (38\text{-}20)$$

Equations 38-17 and 38-20 show that the total energy—and therefore also the kinetic energy—increases without limit as the particle speed v approaches the speed of light c (that is, as v/c approaches 1). And indeed we can add as much kinetic energy as we want to a moving particle in order to increase the energy of collision with other particles, as is done in ever more powerful and ingeniously designed particle accelerators. Yet even the highest-energy particle never moves faster than light as measured in any frame (Fig. 38-7). This result provides the answer to the question asked at the beginning of this chapter: *How can the energy of a particle increase without limit while its speed remains slower than the speed of light?*

Kinetic Energy at Everyday Speeds

In our everyday world the fastest speed we encounter is probably that of a fighter plane moving above the speed of sound at Mach 3 (three times the speed of sound or about 1000 m/s). This speed is not even close to the speed of light. In fact, $v/c = (1000)/(3 \times 10^8 \text{ m/s}) \approx 3 \times 10^{-6}$.

    Equation 38-20 looks complicated. However, for speeds much less than the speed of light (that is, for $v \ll c$) the equation reduces to the Newtonian expression for kinetic energy. To see this, we can use the following approximation:

$$(1 + d)^n \approx 1 + nd \quad \text{for } |d| \ll 1 \text{ and } |nd| \ll 1. \quad (38\text{-}21)$$

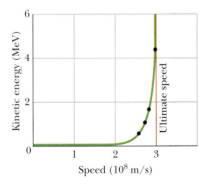

FIGURE 38-7 ■ The dots show measured values of the kinetic energy of an electron plotted against its measured speed. No matter how much energy is given to an electron (or to any other particle having mass), it cannot be accelerated to a speed that equals or exceeds the ultimate limiting speed c. (The curve drawn through the dots shows the predictions of Einstein's special theory of relativity.)

Apply this approximation to the first term on the right of Eq. 38-20. Then in the limit of small values for v^2/c^2, Eq. 38-20 reduces to

$$K = mc^2(1 - v^2/c^2)^{-1/2} - mc^2$$
$$\approx mc^2(1 + v^2/2c^2) - mc^2 \qquad \text{(for } v \ll c). \qquad (38\text{-}22)$$
$$\approx \frac{1}{2}mv^2$$

In brief, the expression for kinetic energy reduces to the Newtonian form for speeds very much less than the speed of light. This limiting case helps to justify our assignment of the name *energy* to the relativistic expressions shown in Eqs. 38-17 and 38-20. The total energy in these equations includes *both* relativistic kinetic energy and rest energy.

Proper Time and Proper Distance in Three Space Dimensions

So far, for simplicity, we have used one space dimension x in the equations of this chapter. Of course, there are three space dimensions, the additional two dimensions often labeled with the symbols y and z. The more general expressions for the time-like and space-like intervals, Eqs. 38-2 and 38-12, are

$$(c\,\Delta\tau)^2 = (c\,\Delta t)^2 - [(\Delta x)^2 + (\Delta y)^2 + (\Delta z)^2] \quad \text{(squared time-like interval),} \quad (38\text{-}23)$$

and

$$(\Delta\sigma)^2 = [(\Delta x)^2 + (\Delta y)^2 + (\Delta z)^2] - (c\,\Delta t)^2 \quad \text{(squared space-like interval).} \quad (38\text{-}24)$$

READING EXERCISE 38-10: Find the speed v at which the energy E of a particle is equal to twice its rest energy. ∎

TOUCHSTONE EXAMPLE 38-6: Energy of a Fast Particle

A particle of mass m moves so fast that its total energy is equal to 1.1 times its rest energy.

(a) What is the speed v of the particle?

SOLUTION ∎ From Eq. 38-19, the rest energy is equal to

$$E_{rest} = mc^2.$$

The statement of the example says we are looking for a speed such that the energy is 1.1 times the rest energy. From Eq. 38-17, we have

$$E = \frac{mc^2}{(1 - v^2/c^2)^{1/2}} = 1.1E_{rest} = 1.1mc^2.$$

Cancel mc^2 from the second and fourth of these equal quantities and equate the results:

$$\frac{1}{(1 - v^2/c^2)^{1/2}} = 1.1.$$

Square both sides and solve the resulting equation for v^2/c^2:

$$v^2/c^2 = 0.1735$$

from which

$$v = 0.416\,c. \qquad \text{(Answer)}$$

(b) What is the kinetic energy of the particle?

SOLUTION ∎ From Eq. 38-20, the kinetic energy is just the total energy minus the rest energy. In our case,

$$K = E - E_{rest} = 1.1E_{rest} - E_{rest} = 0.1E_{rest}. \qquad \text{(Answer)}$$

TOUCHSTONE EXAMPLE 38-7: Decay of K° Meson

A K° meson at rest decays into a π^+ meson and a π^- meson. As is often done in particle physics, we express the masses of these particles in terms of energy divided by c^2. The masses of the two π mesons are identical, so call both m_π. Then

$$m_K = 497.7 \text{ MeV}/c^2$$
$$m_\pi = 139.6 \text{ MeV}/c^2.$$

What is the speed of each of the π mesons after the decay?

SOLUTION ◼ This is assumed to be an isolated system, in which both the total relativistic energy and the total relativistic momentum are conserved. The initial K° meson is at rest, so there is zero momentum before and therefore also after the collision. From this we conclude that the momenta labeled p_1 and p_2 in Fig. 38-8 have equal magnitudes but opposite directions. From Eq. 38-18, this guarantees that the two resulting particles also have equal energies

after the decay. The conservation of energy equates *total* relativistic energy before and after the collision, and this total energy includes rest energy. Before the decay there is only the rest energy of the K° meson. After the decay, there are two π mesons of equal mass and equal energy. From Eq. 38-17,

$$m_K c^2 = 2E_\pi = \frac{2m_\pi c^2}{(1 - v^2/c^2)^{1/2}}.$$

Into the first and last expressions in this equation, substitute the values for the masses from the earlier equations. The factors c^2 cancel, and the units MeV cancel, to yield the equation

$$497.7 = \frac{2 \times 139.6}{(1 - v^2/c^2)^{1/2}}.$$

Square both sides of this equation, solve for v^2/c^2, and take the square root. The answer is

$$v = 0.828\,c. \qquad \text{(Answer)}$$

This is the speed of each π meson as the two move in opposite directions after the decay.

FIGURE 38-8 ◼   BEFORE   AFTER

38-11 The Lorentz Transformation

Most textbooks on special relativity do not emphasize *invariant* quantities that have the same value for all observers. Instead, they focus on the so-called **Lorentz transformation** equations. These equations connect the distinct space and time separations between two events as measured in one frame with those separations as measured in another frame moving past the first. We display the Lorentz transformation equations here without deriving them. (The derivation is not difficult and depends only on the Principle of Relativity and some symmetry arguments.)

In writing the Lorentz transformations it is customary to let unprimed coordinates represent measurements made in the laboratory frame and primed coordinates represent corresponding measurements made in a rocket frame that moves past the laboratory with relative speed v^{rel} along the positive x direction. Then the Lorentz equations that transform the laboratory space and time separations to rocket space and time separations are

$$\Delta x' = \frac{\Delta x - v^{\text{rel}}\Delta t}{(1 - (v^{\text{rel}})^2/c^2)^{1/2}}$$

$$\Delta t' = \frac{\Delta t - (v^{\text{rel}}\Delta x/c^2)}{(1 - (v^{\text{rel}})^2/c^2)^{1/2}} \qquad \text{(Lorentz transformation).} \qquad \text{(38-25)}$$

$$\Delta y' = \Delta y$$

$$\Delta z' = \Delta z$$

What happens to these equations when they describe our everyday life in which typical moving objects are trains, airplanes, and automobiles? These vehicles move at speeds very much less than the speed of light. Assume that v^{rel} is much less than c, so that $v^{rel}/c \ll 1$ in Eqs. 38-25. Then the first two of these equations become

$$\Delta x' = \Delta x - v^{rel}\Delta t$$
$$\Delta t' = \Delta t$$

$\left(v^{rel}/c \ll 1, \text{Galilean transformation}\right).$ (38-26)

These equations are called the **Galilean transformation equations** because they lay out the consequences of relative motion first described by Galileo Galilei in the 1630s. The second equation tells us that for small relative velocities the time between two events has the same value for the moving observer as for the stationary observer. This is certainly typical of our experience; we do not need to reset our watches after an automobile ride! The first equation makes everyday sense as well. A race begins with a starting gun at the starting line and ends with the firing of an "ending gun" at the finish line when the winner crosses it. For observers in the stands the two firings occur a distance Δx and a time Δt apart. Running as fast as we can, we come in second, behind the winner. For us the ending gun goes off a distance ahead of us given by the distance of the racecourse, Δx, minus the distance we have run ($v^{rel}\Delta t$) at speed v^{rel} during the time Δt between the starting and ending guns. This is just what the first Galilean transformation (Eq. 38-26) tells us.

On the other hand, the Lorentz transformation (Eq. 38-25) gives us the rocket (primed) coordinates of an event if we know the laboratory (unprimed) coordinates of that event. But "laboratory" is just a label; it could represent simply another unpowered spaceship. Then the only difference between laboratory and rocket frames is the artificial difference we have given them. The rocket moves in the *positive x* direction with respect to the laboratory, so the laboratory moves in the *negative x* direction with respect to the rocket. It follows that the inverse transformation—the one that gives unprimed laboratory space and time separations in terms of primed rocket space-time separations—can be derived from the Lorentz transformation (Eq. 38-25) merely by interchanging primed and unprimed coordinates and reversing the sign of v^{rel}, making all v-terms positive in the numerators. (Reversing the sign of v^{rel} does not change the sign of $(v^{rel})^2$ in the denominators.) This leads to the equations

$$\Delta x = \frac{\Delta x' + v^{rel}\Delta t'}{(1 - (v^{rel})^2/c^2)^{1/2}}$$

$$\Delta t = \frac{\Delta t' + v^{rel}\Delta x'/c^2}{(1 - (v^{rel})^2/c^2)^{1/2}}$$ (the inverse Lorentz transformation). (38-27)

$$\Delta y = \Delta y'$$

$$\Delta z = \Delta z'$$

38-12 Lorentz Contraction

The Lorentz transformation equations predict a relativistic effect important in the history of the subject—namely, that we as observers will measure an object moving past us at high speed to be shortened—contracted—along its direction of relative motion (but not changed in dimension perpendicular to this direction).

The Lorentz transformation describes the space and time separations between a pair of events. What events can we use to measure the length of a moving object? One choice is the explosions of two firecrackers, one at each end of the object *at the same time* ($\Delta t = 0$) in our frame. Then we can define the length to be the distance Δx between the explosions as measured in our frame. By setting $\Delta t = 0$ in the first term of Eqs. 38-25 and multiplying through by the square-root quantity, we get

$$\sqrt{1 - v^2/c^2}\,\Delta x' = \Delta x.$$ (38-28)

This equation tells us that the length Δx we measure for the object in the laboratory frame—the distance between simultaneous firecracker explosions—is less than the distance $\Delta x'$ between the two ends as measured in the rocket frame in which the object is at rest.

The Lorentz contraction is a curiosity, and is not used very often to analyze experiments. However, it is the consequence of a deeper principle, the relativity of simultaneity, discussed following Eq. 38-12 in Section 38-8 and in Section 38-9. The two firecrackers may explode at the same time in our frame ($\Delta t = 0$) but not in the frame of the rocket ($\Delta t' \neq 0$), as you can see by substituting $\Delta t = 0$ into the second of Eqs. 38-25. The same-time explosions at the two ends of the moving rod in our frame yield a measure of the length of the moving object in our frame. In contrast, the *lack* of simultaneity in the rocket frame does not change the rocket measurement of $\Delta x'$ because the object is at rest in the rocket frame; the distance between the explosions at the two ends is the same whether or not these explosions occur at the same time. However, this lack of simultaneity allows observers in the two frames to account for the difference in measured length of the object.

READING EXERCISE 38-11: What is the speed v of a passing rocket in the case that we measure the length of the rocket to be half its length as measured in a frame in which the rocket is at rest?   ■

38-13 Relativity of Velocities

The speed of light c is the ultimate speed according to special relativity. No object can be accelerated from rest to a speed greater than c. Yet some have found in this statement a paradox that challenges the validity of special relativity. Someone who objects to relativity says, "I ride in a rocket moving at 3/4 the speed of light with respect to the laboratory. From my rocket I launch a stone forward at 3/4 the speed of light as measured in my rocket frame. The result should be an object moving at 3/4 + 3/4 = 1.5 times the speed of light in the laboratory frame. But relativity says that nothing can move faster than light. So my thought experiment shows special relativity to be illogical—and disproves it!"

Since the Lorentz transformation deals with space and time separation between *events*, we can use it to investigate the validity of this thought experiment. Let the stone that you launch forward from your rocket emit two flashes close together. Call the separation between these flash emissions $\Delta x'$ and $\Delta t'$ in your rocket frame and Δx and Δt in our laboratory frame. Then we can derive the velocities of the stone in the two frames from the differential limits of $\Delta x'/\Delta t'$ (velocity in rocket frame) and $\Delta x/\Delta t$ (velocity in laboratory frame).

Next we can use the first two entries in the Lorentz transformation (38-27):

$$\Delta x = \frac{\Delta x' + v^{\text{rel}}\Delta t'}{(1 - (v^{\text{rel}})^2/c^2)^{1/2}}$$

$$\Delta t = \frac{\Delta t' + v^{\text{rel}}\Delta x'/c^2}{(1 - (v^{\text{rel}})^2/c^2)^{1/2}}.$$

Then we can divide corresponding sides of these two equations into each other. The square root expressions in the denominators cancel and we have

$$\frac{\Delta x}{\Delta t} = \frac{\Delta x' + v^{\text{rel}}\Delta t'}{\Delta t' + v^{\text{rel}}\Delta x'/c^2}. \tag{38-29}$$

On the right side of this equation, divide both the numerator and denominator by $\Delta t'$:

$$\frac{\Delta x}{\Delta t} = \frac{\Delta x'/\Delta t' + v^{\text{rel}}}{1 + v^{\text{rel}}(\Delta x'/\Delta t')/c^2}. \tag{38-30}$$

Finally, we can take the differential limit and define u as the velocity of the stone in the laboratory frame and u' as the velocity of the stone in the rocket frame. Then Eq. 38-30 becomes

$$u = \frac{u' + v^{\text{rel}}}{1 + u'v^{\text{rel}}/c^2} \qquad \text{(law of addition of velocities)}. \tag{38-31}$$

Although Eq. 38-31 is called the **law of addition of velocities,** this is not a good name, because the addition is not simple.

How does the law of addition of velocities support the conclusion that nothing can move faster than the speed of light? Let's use this law to find the value of the stone's velocity u in our laboratory frame when the rocket moves away from us at 3/4 the speed of light ($v^{\text{rel}} = 0.75c$) while the stone moves away from the rocket at 3/4 the speed of light ($u' = 0.75c$) as measured in the rocket frame. Substituting these values into Eq. 38-31 yields

$$u = \frac{0.75c + 0.75c}{1 + (0.75c)^2/c^2} = \frac{1.5c}{1 + 0.5625} = 0.96c. \tag{38-32}$$

The result is that we observe the stone to move at the speed $0.96c$ in our laboratory frame. Its speed does not exceed the speed of light. Once again relativity saves itself from disproof!

READING EXERCISE 38-12: A rocket moves with speed $0.9c$ in our laboratory frame. A flash of light is sent forward from the front end of the rocket. Is the speed of that flash equal to $1.9c$ as measured in our laboratory frame? If not, what is the speed of the light flash in our frame? Verify your answer using Eq. 38-31. ∎

TOUCHSTONE EXAMPLE 38-8: Relative Speed of Light

A rocket moves with speed $0.9c$ in our laboratory frame. A flash of light is fired *backward* from the rear of the rocket.

(a) What is the *speed* of that light flash in our laboratory frame?

SOLUTION ■ Using the Principle of Relativity, we can give an answer immediately, without doing any calculations. The speed of light is the same—invariant—in all reference frames. Therefore, the speed of the light flash in our laboratory frame will be c as usual, and in the backward direction. The equations should verify this result.

Use Eq. 38-31:

$$u = \frac{u' + v^{\text{rel}}}{1 + u'v^{\text{rel}}/c^2}.$$

In the example, $v^{\text{rel}} = 0.9c$ and $u' = -c$, with a minus since the light is fired out the back of the rocket:

$$u = \frac{-c + 0.9c}{1 - 0.9c^2/c^2} = \frac{-0.1c}{0.1} = -c. \qquad \text{(Answer)}$$

Therefore the speed of the light flash is equal to c in our laboratory frame, as it has to be and as predicted at the beginning of this solution.

(b) What is the *direction* of the light flash in our laboratory frame?

SOLUTION ■ The minus sign in the most recent equation tells us that in the laboratory frame the light flash moves in a direction opposite to that of the rocket in the laboratory frame.

38-14 Doppler Shift

A car approaches us and passes while sounding its horn. We hear the pitch of the horn decrease as it passes, a change called the **Doppler shift** or **Doppler effect.** In Chapter 18 we found that the Doppler effect depends on two velocities—namely, the velocity of the source and the velocity of the detector with respect to the air, the medium that transmits the waves.

Light and other electromagnetic waves follow a different rule for the Doppler shift, because they require no transmitting medium and can travel through a vacuum. For this reason, the Doppler effect for light depends on only one velocity, the relative velocity between source and detector.

Suppose that a source of light moves away from us with speed v^{rel}, while sending light backward toward us. Let the frequency of the light as measured by the source be denoted f_0. We detect a smaller frequency f than the frequency f_0 and for two reasons: (1) Each wave crest is emitted at a greater distance from us than the previous wave crest, so it has to travel farther to reach us, and (2) for us the traveling clock runs slow (Eq. 38-3). These two effects combine to change the frequency of the light that we receive according to the equation

$$f = f_0 \left[\frac{1 - (v^{rel}/c)}{1 + (v^{rel}/c)} \right]^{1/2} \qquad \text{(source moving away).} \qquad (38\text{-}33)$$

In the case that the source moves *toward* us with speed v^{rel}, we simply reverse the sign of v^{rel}, which occurs twice in Eq. 38-33.

We can measure the frequencies emitted by luminous elements, such as hydrogen excited by an electric discharge in the laboratory. Looking out at stars in nearby galaxies, we can identify hydrogen from the pattern of emitted frequencies. We notice that galaxies farther from us have light shifted downward in frequency compared with their laboratory value and conclude that these galaxies are moving away from us; we call this the **red shift.**

Solving Eq. 38-33 for v^{rel}, we can determine the velocity with which a galaxy recedes from us. This analysis is approximately correct for nearby galaxies. However, the red shift due to the most remote galaxies is not due to the Doppler shift described by special relativity. Rather, the stretching out of the light waves heading toward us occurs because space itself is stretching as the universe expands over time. Thus, general relativity is required to describe this stretching of space with time.

READING EXERCISE 38-13: A not-too-distant galaxy is moving directly away from the Earth. Light from this galaxy includes a pattern of frequencies recognized as those emitted by hydrogen gas. We detect one of these frequencies to have the value $f = 0.9 f_0$, where f_0 is the corresponding frequency for light from hydrogen gas at rest in the laboratory. How fast is the distant galaxy moving away from the Earth? ■

TOUCHSTONE EXAMPLE 38-9: Colliding with Andromeda

According to some predictions, the Andromeda galaxy, currently two million light-years away from us, is moving toward our galaxy and the two will collide in three billion years. Light of a particular frequency f_0 is emitted from hydrogen gas in the stars of the Andromeda galaxy. What is the frequency f of that light measured on Earth?

SOLUTION ■ Equation 38-33 describes the case in which the source is moving directly *away* from us. But in this example, the Andromeda galaxy is moving directly *toward* us. For a source moving directly toward us, reverse the sign of v^{rel} in Eq. 38-33:

$$f = f_0 \left[\frac{1 + (v^{rel}/c)}{1 - (v^{rel}/c)} \right]^{1/2} \qquad \text{(source approaching).}$$

Andromeda is two million light-years distant and is predicted to reach our galaxy in three billion years. Therefore it must be moving at the following fraction of the speed of light,

$$\frac{v^{\text{rel}}}{c} = \frac{2 \times 10^6 \ (\text{light}) \ \text{years}}{3 \times 10^9 \ \text{years}} = 6.667 \times 10^{-4}.$$

This has a magnitude very much less than one, so we can apply approximation Eq. 38-21 to the equation above:

$$f = f_0 \left[\frac{1 + (v^{\text{rel}}/c)}{1 - (v^{\text{rel}}/c)} \right]^{1/2}$$

$$= f_0 \{1 + (v^{\text{rel}}/c)\}^{1/2} \{1 - (v^{\text{rel}}/c)\}^{-1/2}$$

$$\approx f_0 \{1 + (v^{\text{rel}}/2c)\} \{1 + (v^{\text{rel}}/2c)\}$$

$$\approx f_0 \{1 + (v^{\text{rel}}/c) + (v^{\text{rel}}/2c)^2\}.$$

Now, the second term in the last parenthesis has the approximate value 10^{-3}, while the third term has the approximate value 10^{-7}. Therefore we neglect the third term and reach our result:

$$f \approx f_0 \{1 + (v^{\text{rel}}/c)\} = f_0 \{1 + 6.667 \times 10^{-4}\} \approx f_0 \{1 + 7 \times 10^{-4}\}.$$

(Answer)

This expression tells us how much higher is the frequency of light from Andromeda than the frequency of the light from a source of the same atoms viewed at rest in a laboratory on Earth.

Bibliography and Acknowledgments

For a fuller treatment of special relativity that follows the outline presented in this chapter, see *Spacetime Physics*, 2nd Edition, by Edwin F. Taylor and John Archibald Wheeler, W. H. Freeman, New York, 1992, ISBN 0-7167-2327-1. For the relativity of simultaneity, see pages 62–63 and 128–131. For a derivation of the Lorentz transformation equations, see pages 99–103. For a derivation of the Doppler equations, see pages 114 and 263. Some exercises from the Taylor–Wheeler text were adapted for exercises in the present chapter with permission of one author. The present chapter also adapts some excerpts from the entry "Special Relativity" by Edwin F. Taylor in Volume 3 of *The Macmillan Encyclopedia of Physics*, John S. Rigden, Editor in Chief, Simon & Schuster Macmillan, New York, 1996, ISBN 0-02-864588-X.

For a treatment of special relativity that pays more attention to the experimental foundations, see *Special Relativity*, by A. P. French, W. W. Norton, New York, 1968, ISBN 0-393-09793-5. For the relativity of simultaneity, see page 74. For a derivation of the Lorentz transformation equations, see pages 76–82. For a derivation of the Doppler equations, see pages 134–146. Several exercises at the end of Chapter 38 were adapted from the French text with permission of the author.

Albert Einstein's original publication on special relativity is "Zur Elektrodynamik bewegter Körper" ("On the Electrodynamics of Moving Bodies") in *Annalen der Physik*, Volume 17, pages 891–921 (1905). This is reprinted, along with many other original articles on the subject, in the book *The Principle of Relativity* by Albert Einstein, H. A. Lorentz, H. Weyl, H. Minkowski, and others, Dover Books, New York 1952, ISBN 0-486-60081-5.

Arthur I. Miller presents a modern English translation of Einstein's original article, together with a wealth of information on the historical and scientific background and immediate consequences, in his book *Albert Einstein's Special Theory of Relativity: Emergence (1905) and Early Interpretation (1905–1911)*, Addison Wesley, Reading, Massachusetts, 1981, ISBN 0-201-04680-6.

The example of the loud tape player in Reading Exercise 38-9 was devised by Rachel Scherr, Stamatis Vokos, Peter Shaffer, and Andrew Boudreaux.

The author thankfully acknowledges both strategic and detailed suggestions and advice from Patrick J. Cooney, Priscilla W. Laws, and Edward F. Redish. Their help has greatly improved both the physics content and the pedagogical effectiveness of the text, though they are not responsible for any remaining errors.

Problems

SEC. 38-2 ■ ORIGINS OF SPECIAL RELATIVITY

1. Chasing Light. What fraction of the speed of light does each of the following speeds v represent? That is, what is the value of the ratio v/c? (a) A typical rate of continental drift, 3 cm/y. (b) A highway speed limit of 100 km/h. (c) A supersonic plane flying at Mach 2.5 = 3100 km/h. (d) The Earth in orbit around the Sun at 30 km/s. (e) What conclusion(s) do you draw about the need for special relativity to describe and analyze most everyday phenomena? (*Note: Some* everyday phenomena can be derived from relativity. For example, magnetism can be described as arising from electrostatics plus special relativity applied to the slow-moving charges in wires.)

SEC. 38-3 ■ THE PRINCIPLE OF RELATIVITY

2. Fast Computation. A "serial computer," one that carries out one instruction at a time, executes an instruction by transmitting data from the memory to the processor (where computation takes place) and then transmitting the result back to the memory. Estimate the maximum size of a serial "teraflop" computer, one that carries out 10^{12} instructions per second.

3. Examples of the Principle of Relativity. Identical experiments are carried out (1) in a high-speed train moving at constant speed along a horizontal track with the shades drawn and (2) in a closed freight container on the platform as the train passes. Copy the following list and mark with a "yes" quantities that will necessarily be the same as measured in the two frames. Mark with a "no" quantities that are not necessarily the same as measured in the two frames. (a) The time it takes for light to travel one meter in a vacuum; (b) the kinetic energy of an electron accelerated from rest through a voltage difference of one million volts; (c) the time for half the number of radioactive particles at rest to decay; (d) the mass of a proton; (e) the structure of DNA for an amoeba; (f) Newton's Second Law of Motion: $F = ma$; (g) the value of the downward acceleration of gravity g.

4. Riding to Alpha Centauri. You are taking a trip from the solar system to our nearest visible neighbor, Alpha Centauri, approximately 4 light-years distant. At launch you experienced a period of acceleration that increased your speed with respect to Earth from zero to nearly half the speed of light. Now your spaceship is coasting in unpowered flight. Compare and contrast the observations you make now with those you made before the rocket took off from the Earth's surface. Be as specific and detailed as possible. Distinguish between observations made inside the cabin with the windows covered and those made looking out of uncovered windows at the front, side, and back of the cabin.

SEC. 38-4 ■ LOCATING EVENTS WITH AN INTELLIGENT OBSERVER

5. Deducing a Speed. A pulse of protons arrives at detector D, where you are standing. Prior to this, the pulse passed through detector C, which lies 60 meters upstream. Detector C sent a light flash in your direction at the same instant that the pulse passed through it. At detector D you receive the light flash and the proton

pulse separated by a time of 2 nanoseconds (2×10^{-9} s). What is the speed of the proton pulse?

6. Eruption from the Sun. You see a sudden eruption on the surface of the Sun. From solar theory you predict that the eruption emitted a pulse of particles that is moving toward the Earth at one-eighth the speed of light. How long do you have to seek shelter from the radiation that will be emitted when the particle pulse hits the Earth? Take the light-travel time from the Sun to the Earth to be 8 minutes.

SEC. 38-5 ■ LABORATORY AND ROCKET LATTICEWORKS OF CLOCKS

7. Synchronizing a Clock. In a vast latticework of meter sticks and clocks, you stand next to a lattice clock whose coordinates are $x = 8$ km, $y = 40$ km, $z = 44$ km. When you receive the synchronizing flash, to what time do you quickly set your clock?

8. Earth's Surface Inertial? Quite apart from effects due to the Earth's rotational and orbital motion, a laboratory reference frame on the Earth is not an inertial frame, as required by a strict interpretation of special relativity. It is not inertial because a particle released from rest at the Earth's surface does not remain at rest; it falls! Often, however, the events in an experiment for which one needs special relativity happen so quickly that we can ignore effects due to gravitational acceleration. Consider, for example, a proton moving horizontally at speed $v = 0.992c$ through a 10-m-wide detector in a laboratory test chamber. (a) How long will the transit through that detector take? (b) How far does the proton fall vertically during this time lapse? (c) What do you conclude about the suitability of the laboratory as an inertial frame in this case?

SEC. 38-6 ■ TIME STRETCHING

9. Light Clock for a Faster Rocket. Redo Fig. 38-5 with a vertical distance $c\,\Delta\tau/2 = 7$ m and horizontal distance in the lab frame $\Delta x/2 = 24$ m. Find the ratio of the times $\Delta t/\Delta\tau$ between events A and B recorded on laboratory and rocket clocks.

SEC. 38-7 ■ THE METRIC EQUATION

10. Where and When? Two firecrackers explode at the same place in the laboratory and are separated by a time of 12 years. (a) What is the spatial distance between these two events in a rocket in which the events are separated in time by 13 years? (b) What is the relative speed of the rocket and laboratory frames? Express your answer as a fraction of the speed of light.

11. Traveling to Vega. Jocelyn DeGuia takes off from Earth and moves toward the star Vega, which is 26 ly distant from Earth. Assume that Earth and Vega are relatively at rest and Jocelyn moves at $v = 0.99c$ in the Earth–Vega frame. How much time will have elapsed on Earth (a) when Jocelyn reaches Vega and (b) when Earth observers receive a radio signal reporting that Jocelyn has arrived? (c) How much will Jocelyn age during her outward trip?

12. Travel to the Dog Star. In the 24th century the fastest available interstellar rocket moves at $v = 0.75c$. Mya Allen is sent in this rocket at full (constant) speed to Sirius, the Dog Star, the brightest star in the heavens as seen from Earth, which is a distance 8.7 ly as measured in the Earth frame. Assume Sirius is at rest with respect to Earth. Mya stays near Sirius, slowly orbiting around that Dog Star, for 7 years as recorded on her wristwatch while making observations and recording data, then returns to Earth with the same speed $v = 0.75c$. According to Earth-linked observers: (a) When does Mya arrive at Sirius? (b) When does Mya leave Sirius? (c) When does Mya arrive back at Earth? According to Mya's wristwatch: (d) When does she arrive at Sirius? (e) When does she leave Sirius? (f) When does she arrive back on Earth?

13. Fast-Moving Muons. The half-life of stationary muons is measured to be 1.6 microseconds. Half of any initial number of stationary muons decays in one half-life. Cosmic rays colliding with atoms in the upper atmosphere of the Earth create muons, some of which move downward toward the Earth's surface. The mean lifetime of high-speed muons in one such burst is measured to be 16 microseconds. (a) Find the speed of these muons relative to the Earth. (b) Moving at this speed, how far will the muons move in one half-life? (c) How far would this pulse move in one half-life if there were no relativistic time stretching? (d) In the relativistic case, how far will the pulse move in 10 half-lives? (e) An initial pulse consisting of 10^8 muons is created at a distance above the Earth's surface given in part (d). How many will remain at the Earth's surface? Assume that the pulse moves vertically downward and none are lost to collisions. (Ninety-nine percent of the Earth's atmosphere lies below 40 km altitude.)

14. Lifetime of a Fast Particle. An unstable high-energy particle is created in a collision inside a detector and leaves a track 1.05 mm long before it decays while still in the detector. Its speed relative to the detector was $0.992c$. How long did the particle live as recorded in its rest frame?

15. Living a Thousand Years in One Year. You wish to make a round trip from Earth in a spaceship, traveling at constant speed in a straight line for 6 months on your watch and then returning at the same constant speed. You wish, further, to find Earth to be 1000 years older on your return. (a) What is the value of your constant speed with respect to Earth? (b) How much do you age during the trip? (c) Does it matter whether or not you travel in a straight line? For example, could you travel in a huge circle that loops back to Earth?

16. Birthdays. An astronaut traveling in an unpowered spaceship celebrates his 18th, 19th, 20th, and 21st birthdays. Five Earth-years elapse between the 18th and 21st birthday parties. Find (a) the spatial separation between the 18th and 21st birthday parties in the Earth frame and (b) the speed of his spaceship with respect to Earth.

SEC. 38-8 ■ CAUSE AND EFFECT

17. Relations Between Events. The table shows the t and x coordinates of three events as observed in the laboratory frame.

Laboratory Coordinates of Three Events

Event	t years	x light-years
Event 1	2	1
Event 2	7	4
Event 3	5	6

On a piece of paper list vertically every pair of these events: (1, 2), (1, 3), (2, 3). (a) Next to each pair write "time-like," "light-like," or "space-like" for the relationship between those two events. (b) Next to each pair, write "Yes" if it is possible for one of the events to cause the other event and "No" if a cause and effect relation between them is not possible. (For full benefit of this exercise, construct and analyze your own tables.)

18. Proper Distance and Proper Time. Use the equations in Chapter 38 to show the following general results: (a) Given that two events P and Q have a space-like separation, show that in all such cases a reference frame can be found in which the two events occur at the same time. Also show that with respect to this frame the distance between the two events is equal to the proper distance between them. (b) Given that two events P and R have a time-like separation, show that in all such cases a reference frame can be found in which the two events occur at the same place. Also show that in this frame the time lapse between the two events is equal to the proper time between them. (c) Given that two events R and W have a light-like separation, show that in all such cases a light flash can be found that moves from R to W. Also show that the proper time and proper distance between R and W are both equal to zero.

SEC. 38-9 ■ RELATIVITY OF SIMULTANEITY

19. Symmetric Relativity of Simultaneity. In the thought experiment pictured in Fig. 38-6, we arbitrarily chose events so that the two light flashes from the lightning strikes arrived simultaneously at the ground observer. Analyze a new version of this experiment in which a completely different pair of lightning strikes fall at the two ends of the train such that the resulting light flashes arrive simultaneously at the position of the rider at the center of the train. View the experiment in the rest frame of the train. In this new version of the experiment, which lightning bolt falls first according to the observer on the ground?

SEC. 38-10 ■ MOMENTUM AND ENERGY

20. Boosting the Speed. How much work must be done to increase the speed of an electron (a) from $0.08c$ to $0.09c$? (b) from $0.98c$ to $0.99c$? Note that the increase in speed is the same in both cases.

21. Lightbulb Radiating Mass. How much mass does a 100 W lightbulb dissipate (in heat and light) when it burns for one full year?

22. Proton Crosses Galaxy. Find the energy of a proton that crosses our galaxy (diameter 100 000 light-years) in one minute of its own time.

23. Converting Mass to Energy. The values of the masses in the reaction

$$p + {}^{19}F \rightarrow \alpha + {}^{16}O$$

have been determined by a mass spectrometer to have the values:

$$m(p) = 1.007825u,$$
$$m(F) = 18.998405u,$$
$$m(\alpha) = 4.002603u,$$
$$m(O) = 15.994915u.$$

Here u is the atomic mass unit (Section 1.7). How much energy is released in this reaction? Express your answer in both kilograms and MeV.

24. Aspirin-Powered Automobile. An aspirin tablet contains 5 grains of aspirin (medicinal unit), which is equal to 325 mg. For how many kilometers would the energy equivalent of this mass power an automobile? Assume 12.75 km/L and a heat of combustion of 3.65×10^7 J/L for the gasoline used in the automobile.

25. Converting Energy to Mass. Two freight trains, each of mass 6×10^6 kg (6 000 metric tons) travel in opposite directions on the same track with equal speeds of 150 km/hr. They collide head-on and come to rest. (a) Calculate in joules the kinetic energy $(1/2)mv^2$ for each train before the collision. (Newtonian expression OK for everyday speeds!) (b) After the collision, the mass of the trains plus the mass of the track plus the mass of the roadbed plus the mass of the surrounding air plus the mass of emitted sound and light has increased by what number of milligrams?

26. Electrically Accelerated Electron. Through what voltage must an electron be accelerated from rest in order to increase its energy to 101% of its rest energy?

27. Powerful Proton. A proton exits an accelerator with a kinetic energy equal to N times its rest energy. Find expressions for its (a) speed and (b) momentum.

28. Relativistic Chemistry. One kilogram of hydrogen combines chemically with 8 kilograms of oxygen to form water; about 10^8 J of energy is released. Ten metric tons (10^4 kg) of hydrogen combines with oxygen to produce water. (a) Does the resulting water have a greater or less mass than the original hydrogen plus oxygen? (b) What is the numerical magnitude of this difference in mass? (c) A smaller amount of hydrogen and oxygen is weighed, then combined to form water, which is weighed again. A very good chemical balance is able to detect a fractional change in mass of 1 part in 10^8. By what factor is this sensitivity more than enough—or insufficient—to detect the fractional change in mass in this reaction?

29. Finding the Mass. (a) Find an equation for the unknown mass m of a particle if you know its momentum p and its kinetic energy K. Show that this expression reduces to an expected result for non-relativistic particle speeds. (b) Find the mass of a particle whose kinetic energy is $K = 55.0$ MeV and whose momentum is $p = 121$ MeV/c. Express your answer as a decimal fraction or multiple of the mass m_e of the electron.

30. A Box of Light. Estimate the power in kilowatts used to light a city of 8 million inhabitants. If all this light generated during one hour in the evening could be captured and put in a box, how much would the mass of the box increase?

31. Creating a Proton–Antiproton Pair. Two protons, each of mass m, are fired toward one another with equal energy (see Fig. 38-9). They collide and create an additional proton–antiproton pair, each with the proton mass m. (a) Show that the lowest total energy E of the incident protons for this creation to take place leaves the resulting four particles at rest with respect to one another. The value of this minimum energy for each incident particle is called the **threshold energy.** (b) What is the threshold *kinetic* energy K of each incident particle for this creation to occur? Express your answer in terms of the rest energy of the proton. (c)

FIGURE 38-9 ■ Problem 31.

Given that the mass of a proton is approximately equal to 1 GeV/c^2, what is the value of the threshold kinetic energy of each incident proton? Explain why this result is reasonable.

SEC. 38-11 ■ THE LORENTZ TRANSFORMATION

32. Really Simultaneous? (a) Two events occur at the same time in the laboratory frame and at the laboratory coordinates ($x_1 = 10$ km, $y_1 = 4$ km, $z_1 = 6$ km) and ($x_2 = 10$ km, $y_2 = 7$ km, $z_2 = -10$ km). Will these two events be simultaneous in a rocket frame moving with speed $v^{\text{rel}} = 0.8c$ in the x direction in the laboratory frame? Explain your answer. (b) Three events occur at the same time in the laboratory frame and at the laboratory coordinates (x_0, y_1, z_1), (x_0, y_2, z_2), and (x_0, y_3, z_3), where x_0 has the same value for all three events. Will these three events be simultaneous in a rocket frame moving with speed v^{rel} in the laboratory x direction? Explain your answer. (c) Use your results of parts (a) and (b) to make a general statement about simultaneity of events in laboratory and rocket frames.

33. Transformation of y-velocity. A particle moves with uniform speed $v_y' = \Delta y'/\Delta t'$ in the y' direction with respect to a rocket frame that moves along the x axis of a laboratory frame. Find expressions for the x-component and for the y-component of the particle's velocity in the laboratory frame.

34. Transformation of Velocity Direction. A particle moves with speed v' in the $x'y'$ plane of the rocket frame and in a direction that makes an angle ϕ' with the x' axis. Find the angle ϕ that the velocity vector of this particle makes with the x axis of the laboratory frame. (*Hint:* Transform space and time displacements rather than velocities.)

35. The Headlight Effect. A flash of light is emitted at an angle ϕ' with respect to the x' axis of the rocket frame. (a) Show that the angle ϕ the direction of motion of this flash makes with respect to the x axis of the laboratory frame is given by the equation

$$\cos\phi = \frac{\cos\phi' + v^{\text{rel}}/c}{1 + (v^{\text{rel}}/c)\cos\phi'}.$$

Optional: Show that your answer to Problem 34 gives the same result when the velocity v' is given the value c. (b) A light source at rest in the rocket frame emits light uniformly in all directions. In the rocket frame 50% of this light goes into the forward hemisphere of a sphere surrounding the source. Show that in the laboratory frame this 50% of the light is concentrated in a narrow forward cone of half-angle ϕ_0 whose axis lies along the direction of motion of the particle. Derive the following expression for the half-angle ϕ_0:

$$\cos\phi_0 = v^{\text{rel}}/c.$$

This result is called the **headlight effect.** (c) What is the half-angle ϕ_0 in degrees for a light source moving at 99% of the speed of light?

SEC. 38-12 ■ LORENTZ CONTRACTION

36. Electron Shrinks Distance. An evacuated tube at rest in the laboratory has a length 3.00 m as measured in the laboratory. An electron moves at speed $v = 0.999\,987c$ in the laboratory along the axis of this evacuated tube. What is the length of the tube measured in the rest frame of the electron?

37. Passing Time. A spaceship of rest length 100 m passes a laboratory timing station in 0.2 microseconds measured on the timing station clock. (a) What is the speed of the spaceship in the laboratory frame? (b) What is the Lorentz-contracted length of the spaceship in the laboratory frame?

38. Transformation of Angles. A meter stick lies at rest in the rocket frame and makes an angle ϕ' with the x' axis as measured by the rocket observer. The laboratory observer measures the x- and y-components of the meter stick as it streaks past. From these components the laboratory observer computes the angle ϕ that the stick makes with his x axis. (a) Find an expression for the angle ϕ in terms of the angle ϕ' and the relative speed v^{rel} between rocket and laboratory frames. (b) What is the length of the "meter" stick measured by the laboratory observer? (c) *Optional:* Why is your expression in part (a) different from equations derived in Problems 34 and 35?

39. Traveling to the Galactic Center. (a) Can a person, in principle, travel from Earth to the center of our galaxy, which is 23 000 ly distant, in one lifetime? Explain using either length contraction or time dilation arguments. (b) What constant speed with respect to the galaxy is required to make the trip in 30 y of the traveler's lifetime?

40. Limo in the Garage. Carman has just purchased the world's longest stretch limo, which has proper length $L_c = 30.0$ m. Part (a) of Figure 38-10 shows the limo parked at rest in front of a garage of proper length $L_g = 6.00$ m, which has front and back doors. Looking at the limo parked in front of the garage, Carman says there is no way that the limo can fit into the garage. "*Au contraire!*" shouts Garageman, "Under the right circumstances the limo can fit into the garage with both garage doors closed and room to spare!" Garageman envisions a fast-moving limo that takes up exactly one-third of the proper length of the garage. Part (b) of Figure 38-10 shows the speeding limo just as the front

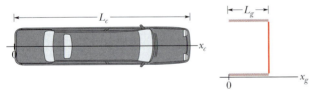

(a) Stretch limo and garage, both at rest in frame of garage.

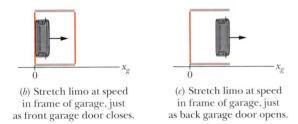

(b) Stretch limo at speed in frame of garage, just as front garage door closes.

(c) Stretch limo at speed in frame of garage, just as back garage door opens.

FIGURE 38-10 ■ Problem 40.

garage door closes behind it as recorded in the garage frame. Part (c) of Figure 38-10 shows the limo just as the back garage door opens in front of it as recorded in the garage frame. Find the speed of the limo with respect to the garage required for this scenario to take place.

SEC. 38-13 ■ RELATIVITY OF VELOCITIES

41. Backfire. An unpowered rocket moves past you in the positive x direction at speed $v^{rel} = 0.9c$. This rocket fires a bullet out the back that you measure to be moving at speed $v_{bullet} = 0.3c$ in the positive x direction. With what speed relative to the rocket did the rocket observer fire the bullet out the back of her ship?

42. Separating Galaxies. Galaxy A is measured to be receding from us on Earth with a speed of $0.3c$. Galaxy B, located in precisely the opposite direction, is also receding from us at the same speed. What recessional velocity will an observer on galaxy A measure (a) for our galaxy, and (b) for galaxy B?

43. Decaying K^o Meson. Touchstone Example 38-7 concluded that when a K^o meson at rest decays into two daughter π mesons, they move in opposite directions in the rest frame of the original K^o meson, each with a speed of $0.828c$. Now suppose that the initial K^o meson moves with speed $v^{rel} = 0.9c$ as measured in the laboratory frame. What are the maximum and minimum speeds of the daughter π mesons with respect to the laboratory?

44. Transit Time. An unpowered spaceship whose rest length is 350 meters has a speed $0.82c$ with respect to Earth. A micrometeorite, also with speed of $0.82c$ with respect to Earth, passes the spaceship on an antiparallel track that is moving in the opposite direction. How long does it take the micrometeorite to pass the spaceship as measured on the ship?

SEC. 38-14 ■ DOPPLER SHIFT

45. Listening to the Traveler. A spaceship moving away from Earth at a speed $0.900c$ radios its reports back to Earth using a frequency of 100 MHz measured in the spaceship frame. To what frequency must Earth's receivers be tuned in order to receive the reports?

46. Speed Trap. How fast would you have to approach a red traffic light in order that it appears green to you?

47. Redshift Factor z. Astrophysicists describe the redshift of receding astronomical objects using the **redshift factor z,** defined implicitly in the following equation:

$$\lambda_{observed} \equiv (1 + z)\lambda_{emitted}.$$

Here $\lambda_{observed}$ is the wavelength of light observed from Earth, while $\lambda_{emitted}$ is the wavelength of the light emitted from the source as measured in the rest frame of the source. The emitted wavelength is known if one knows the emitting atom, identified from the pattern of different wavelengths characteristic of that atom. Astrophysicists measuring the redshifts of light from extremely remote quasars calculate a z-factor in the neighborhood of $z \approx 6$. Use the Doppler shift equations of special relativity to determine how fast such quasars are moving away from Earth. *Note:* Actually, for such distant objects the unmodified Doppler shift formula of special relativity does not apply. Instead, one thinks of the space between Earth and the source expanding as the universe expands; the wavelength of the light expands with this expansion of the universe as it travels from the source quasar to us.

48. Receding Galaxy. Figure 38-11 shows a graph of intensity versus wavelength for light reaching Earth from galaxy NGC 7319, which is about 3×10^8 light-years away. The most intense light is emitted by the oxygen in that galaxy. In a laboratory, that emission is at wavelength $\lambda = 513$ nm, but in the light from NGC 7319 it has been shifted to 525 nm due to the Doppler effect. (Indeed, all the emissions from that galaxy have been shifted.) (a) According to special relativity Doppler shift theory, what is the radial speed of galaxy NGC 7319 relative to Earth? (b) Is the relative motion toward or away from Earth?

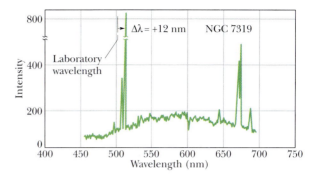

FIGURE 38-11 ■ Problem 48

Additional Problems

49. Exodus from Earth. A billion years from now our Sun will increase its heat, destroying life on Earth. Still later the sun will expand as a red giant, swallowing the Earth and annihilating any remaining life on all planets in the solar system. In anticipation of these catastrophes, an advanced Earth civilization a million years from now develops a transporter mechanism that reduces living beings to data and sends the data by radio to planets orbiting younger stars. The living beings on Earth are destroyed by this process but are reconstituted and restored to life on the distant planets. Your descendent Rasmia Kirmani leaves Earth as data at a time we will take to be zero and is quickly reconstituted after arrival of her data set on the planet Zircon, 100 ly distant from Earth. Assume that Earth and Zircon are relatively at rest.

(a) How much does Rasmia age during her outward trip to Zircon?
(b) How much older is Earth and its civilization when Rasmia is resurrected on Zircon?
(c) Rasmia has a productive and happy life on Zircon and dies as a pioneer hero after 150 years living on that planet. How soon after her departure from Earth can Rasmia's obituary be received on Earth?
(d) Over the millennia between our time and then, specialists whom we now call geneticists discover that there is no such thing as a superperson (man or woman), but rather that a minimum variety of genetic types must be maintained and continually recombined (by whatever method is then current) in order to sustain a healthy population. To this end, several dozen healthy individuals are deconstructed on Earth and transported to Zircon, where each individual is quickly reproduced in thousands of copies (using the same data set over and over) in order to populate the planet

rapidly. It takes 5 full generations from birth to death, each generation an average of 200 years, to determine whether or not the new population has been successfully established. How soon after transmission of the dozens of original data sets from Earth can Earth's people learn whether or not this project has been successful?

50. Electron in Orbit. Use Newtonian mechanics to calculate the speed of an electron in the lowest Bohr orbit, which has one quantum of angular momentum:

$$mvr = \hbar = \frac{h}{2\pi}.$$

Carry out this calculation for (a) hydrogen ($Z = 1$) and (b) uranium ($Z = 92$). Insofar as the Bohr model of the atom can be trusted, is relativity required to find the correct answer for (c) hydrogen, (d) uranium?

51. Super Cosmic Rays. The Giant Shower Array detector, spread over 100 square kilometers in Japan, detects pulses of particles from cosmic rays. Each detected pulse is assumed to originate in a single high-energy cosmic proton that strikes the top of the Earth's atmosphere. The highest energy of a single cosmic ray proton inferred from the data is 10^{20} eV. How long would it take that proton to cross our galaxy (10^5 light-years in diameter) as recorded on the wristwatch of the proton? (The answer is not zero!)

52. Synchronization by a Traveling Clock. Evelyn Brown does not approve of our latticework of rods and clocks and the use of a light flash to synchronize them.

(a) "I can synchronize my clocks in any way I choose!" she exclaims. Is she right?

(b) Evelyn wants to synchronize two identical clocks, called Big Ben and Little Ben, which are at rest with respect to one another and separated by one million kilometers in their rest frame. She uses a third clock, identical in construction with the first two, that travels with constant velocity between them. As her moving clock passes Big Ben, it is set to read the same time as Big Ben. When the moving clock passes Little Ben, that outpost clock is set to read the same time as the traveling clock. "Now Big Ben and Little Ben are synchronized," says Evelyn Brown. Is Evelyn's method correct?

(c) After Evelyn completes her synchronization of Little Ben by her method, how does the reading of Little Ben compare with the reading of a nearby clock on a latticework at rest with respect to Big Ben (and Little Ben) and synchronized by our standard method using a light flash? Evaluate in milliseconds any difference between the reading on Little Ben and the nearby lattice clock in the case that Evelyn's traveling clock moved at a constant velocity of 500 000 kilometers per hour from Big Ben to Little Ben.

(d) Evaluate the difference in the reading between the Evelyn-Brown-synchronized Little Ben and the nearby lattice clock when Evelyn's synchronizing traveling clock moves 1000 times as fast as the speed given in part (c).

53. Down with Relativity! Sara Settlemyer is an intelligent layperson who carefully reads articles about science in the public press. She has the objections to relativity listed below. Respond to each of Sara's objections clearly, decisively, and politely—without criticizing her!

(a) "Observer A says that observer B's clock runs slow, while B says that A's clock runs slow. This is a logical contradiction. Therefore relativity should be abandoned."

(b) "Observer A says that B's meter sticks are contracted along their direction of relative motion. B says that A's meter sticks are contracted. This is a logical contradiction. Therefore relativity should be abandoned."

(c) "Anybody with common sense knows that travel at high speed in the direction of a receding light pulse decreases the speed with which the pulse recedes. Hence a flash of light cannot have the same speed for observers in relative motion. With this disproof of the Principle of Relativity, all of relativity collapses."

(d) "Relativity is preoccupied with how we *observe* things, not with what is *really* happening. Therefore relativity is not a scientific theory, since science deals with *reality*."

(e) "Relativity offers no way to describe an event without coordinates, and no way to speak about coordinates without referring to one or another particular reference frame. However, physical events have an existence independent of all choice of coordinates and reference frames. Therefore the special relativity you talk about in this chapter cannot be the most fundamental theory of events and the relation between events."

54. The Photon as a Zero-Mass Particle. A **photon,** the quantum of light, can be considered to be a zero-mass particle.

(a) Using this definition and Eq. 38-18, show that the relation between energy and momentum for the photon is $E = |pc|$, where the "absolute value" vertical lines ensure that energy is positive.

(b) A π^0 meson decays rapidly into two gamma rays (high-energy photons). In the rest frame of the original π^0 meson, what are the relative directions of the two outgoing photons?

(c) If the mass of the π^0 meson is 135 MeV/c^2, what is the energy of each outgoing gamma ray?

55. Pair Production with Gamma Rays. Two gamma rays of equal energy E_p and equal and opposite momenta are incident on a nucleus. (See Figure 38-12.) The collision leads to annihilation of the gamma rays and creation of an electron–positron pair. The lowest energy (the "threshold energy") of incident photons for this production leaves the resulting electron and positron at rest with respect to the nucleus. (The nucleus acts as midwife to this birth and is not changed by the interaction.)

(a) What is the threshold energy E_p of each photon for this creation to take place?

(b) Generalize Eq. 38-18 to *define* the mass M_s of a system of particles, given the total energy E_s and net momentum p_s of the system:

$$M_s^2 c^4 \equiv E_s^2 - p_s^2 c^2.$$

What is the mass M_s of the system of particles after the collision? Before the collision?

(c) *Mass without mass?* Now let the "nuclear" mass m become less and less. In the limit $m \rightarrow 0$, what is the mass of the system after the collision? Before the collision? Before the collision, you apparently have a system with mass composed of "particles," each of which has zero mass. Does this make sense?

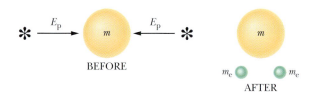

FIGURE 38-12 ■ Problem 55.

56. Resonant Absorption of a Gamma Ray. A gamma ray (an energetic photon) falls on a nucleus of initial mass m, initially at rest. The energy E_p of the incoming gamma ray matches the energy separation between the lowest energy of the nucleus and its first excited state, so the incident photon is absorbed. We want to know the mass m^* of the excited nucleus. (see Fig. 38-13.)

(a) Show that the conservation of energy and momentum equations are, in an obvious notation:

$$E_p + mc^2 = E_{m^*}$$

and

$$\frac{E_p}{c} = p_{m^*} = \frac{(E_{m^*}^2 - m^{*2}c^4)^{1/2}}{c}.$$

(b) Combine the two conservation equations to find an expression for m^* as a function of E_p, m, and c.
(c) Show that for very small values of E_p the limiting result is $m^* = m$. Explain why this limiting result is reasonable.

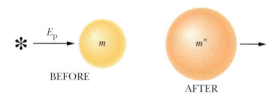

FIGURE 38-13 ■ Problem 56.

57. Photon Braking. A radioactive nucleus of known initial mass M and known initial total energy E_M emits a gamma ray (high-energy photon) in the direction of its motion, drops to its stable nonradioactive state of known mass m, and comes to rest. (see Fig. 38-14). Find an expression for the total energy E_M of the incoming nucleus. The unknown energy E_p of the outgoing gamma ray should not appear in your expression.

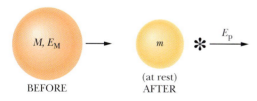

FIGURE 38-14 ■ Problem 57.

58. Limo and Garage Paradox. Review Problem 40, in which we concluded that a limo of proper length 30 m can fit into a garage of proper length 6 m with room to spare. This result is possible because the speeding limo is observed by Garageman to be Lorentz -

contracted. Carman protests that in the rest frame of the limo (in which the limo is its full proper length) it is the *garage* that is Lorentz-contracted. As a result, he claims, there is no possibility whatever that the limo can fit into the garage. What could be the possible basis for resolving this paradox? (*Hint:* Think about the space and time locations of two events: event A, front garage door closes and event B, rear garage door opens.)

59. Twin Paradox. The famous **twin paradox** is often introduced as follows: Two identical twins grow up together on Earth. When they reach adulthood, one twin zooms to a distant star and returns to find her stay-at-home sister much older than she is. Thus far no paradox. But Alexis Allen formulates the Twin Paradox for us:

"The theory of special relativity tells us that all motion is relative. With respect to the traveling twin, the Earth-bound twin moves away and then returns. Therefore it is the Earth-bound twin who should be younger than the 'traveling' twin. But when they meet again at the same place, it cannot possibly be that each twin is younger than the other twin. This Twin Paradox disproves relativity."

The paradox is usually resolved by realizing that the traveling twin turns around. Everyone agrees which twin turns around, since the reversal of direction slams the poor traveler against the bulkhead of the decelerating starship, breaking her collarbone. The turnaround, evidenced by the broken collarbone, destroys the symmetry required for the paradox to hold. Good-bye Twin Paradox!

Still, Alexis's father Cyril Allen has his doubts about this resolution of the paradox. "Your solution is extremely unsatisfying. It forces me to ask: What if the retro-rockets malfunction and will not fire at all to slow me down as I approach a distant star a thousand light-years from Earth? Then I cannot even stop at that star, much less turn around and head back to Earth. Instead, I continue moving away from Earth forever at the original constant speed. Does this mean that as I pass the distant star, one thousand light-years from Earth, it is no longer possible to say that I have aged less than my Earth-bound twin? But if not, then I would never have even gotten to the distant star at all during my hundred-year lifetime! Your resolution of the Twin Paradox is insufficient and unsatisfying."

Write a half-page response to Cyril Allen, answering his objections politely but decisively.

60. The Runner on the Train Paradox. A train moves at 10 km/h along the track. A passenger sprints toward the rear of the train at 10 km/h with respect to the train. Our knee-jerk motto says that the train clocks "run slow" with respect to clocks on the track, and the runner's watch "runs slow" with respect to train clocks. Therefore the runner's watch should "run doubly slow" with respect to clocks on the track. But the runner is at rest with respect to the track. What gives? (This example illustrates the danger of the simple knee-jerk motto "Moving clocks run slow.")

61. A Summer Evening's Fantasy. You and a group of female and male friends stand outdoors at dusk watching the Sun set and noticing the planet Venus in the same direction as the Sun. An alien ship lands beside you at the same instant that you see the Sun explode. The aliens admit that earlier they shot a laser flash at the Sun, which caused the explosion. They warn that the Sun's explosion emitted an immense pulse of particles that will blow away Earth's atmosphere. In confirmation, a short time after the aliens land you notice Venus suddenly change color. You and your friends plead with the aliens to take your group away from Earth in order to establish the human gene pool elsewhere. They agree. Describe the conditions under which your escape plan will succeed. Be specific and use numbers. Assume that the Sun is 8 light-minutes from Earth and Venus is 2 light-minutes from Earth.

The International System of Units (SI)*

1 SI Base Units

1. The SI Base Units

Quantity	Name	Symbol	Definition
length	meter	m	"... the length of the path traveled by light in vacuum in 1/299 792 458 of a second." (1983)
mass	kilogram	kg	"... this prototype [a certain platinum–iridium cylinder] shall henceforth be considered to be the unit of mass." (1889)
time	second	s	"... the duration of 9 192 631 770 periods of the radiation corresponding to the transition between the two hyperfine levels of the ground state of the cesium-133 atom." (1967)
electric current	ampere	A	"... that constant current which, if maintained in two straight parallel conductors of infinite length, of negligible circular cross section, and placed 1 meter apart in vacuum, would produce between these conductors a force equal to 2×10^{-7} newton per meter of length." (1946)
thermodynamic temperature	kelvin	K	"... the fraction 1/273.16 of the thermodynamic temperature of the triple point of water." (1967)
amount of substance	mole	mol	"... the amount of substance of a system which contains as many elementary entities as there are atoms in 0.012 kilogram of carbon-12." (1971)
luminous intensity	candela	cd	"... the luminous intensity, in a given direction, of a source that emits monochromatic radiation of frequency 540×10^{12} hertz and that has a radiant intensity in that direction of 1/683 watt per steradian." (1979)

2 The SI Supplementary Units

2. The SI Supplementary Units

Quantity	Name of Unit	Symbol
plane angle	radian	rad
solid angle	steradian	sr

*Adapted from "The International System of Units (SI)," National Bureau of Standards Special Publication 330, 2001 edition. The definitions above were adopted by the General Conference of Weights and Measures, an international body, on the dates shown. In this book we do not use the candela.

3 Some SI Derivations

3. Some SI Derived Units

Quantity	Name of Unit	Symbol	In Terms of other SI Units
area	square meter	m^2	
volume	cubic meter	m^3	
frequency	hertz	Hz	s^{-1}
mass density (density)	kilogram per cubic meter	kg/m^3	
speed, velocity	meter per second	m/s	
rotational velocity	radian per second	rad/s	
acceleration	meter per second per second	m/s^2	
rotational acceleration	radian per second per second	rad/s^2	
force	newton	N	$kg \cdot m/s^2$
pressure	pascal	Pa	N/m^2
work, energy, quantity of heat	joule	J	$N \cdot m$
power	watt	W	J/s
quantity of electric charge	coulomb	C	$A \cdot s$
potential difference, electromotive force	volt	V	W/A
electric field strength	volt per meter (or newton per coulomb)	V/m	N/C
electric resistance	ohm	Ω	V/A
capacitance	farad	F	$A \cdot s/V$
magnetic flux	weber	Wb	$V \cdot s$
inductance	henry	H	$V \cdot s/A$
magnetic flux density	tesla	T	Wb/m^2
magnetic field strength	ampere per meter	A/m	
entropy	joule per kelvin	J/K	
specific heat	joule per kilogram kelvin	$J/(kg \cdot K)$	
thermal conductivity	watt per meter kelvin	$W/(m \cdot K)$	
radiant intensity	watt per steradian	W/sr	

4 Mathematical Notation

Poorly chosen mathematical notation can be a source of considerable confusion to those trying to learn and to do physics. For example, ambiguity in the meaning of a mathematical symbol can prevent a reader from understanding the meaning of a crucial relationship. It is also difficult to solve problems when the symbols used ot represent different quantities are not distinctive. In this text we have taken special care to use mathematical notation in ways that allow important distinctions to be easily visible both on the printed page and in handwritten work.

An excellent starting point for clear mathematical notation is the U.S. National Institute of Standard and Technology's Special Publication 811 (SP 811), *Guide for the Use of the International System of Units (SI)*, available at http://physics.nist.gov/cuu/Units/bibliography.html. In addition to following the National Institute guidelines, we have made a number of systematic choices to facilitate the translation of printed notation into handwritten mathematics. For example:

- Instead of making vectors bold, vector quantities (even in one dimension) are denoted by an arrow above the symbol. So printed equations look like handwritten equations. Example: $\vec{v}$ rather than v is used to denote an instantaneous velocity.

- In general, each vector component has an explicit subscript denoting that it represents the component along a chosen coordinate axis. The one exception is the position vector, $\vec{r}$, whose components are simply written as x, y, and z. For example, $\vec{r} = x\hat{i} + y\hat{j} + z\hat{k}$, whereas, $\vec{v} = v_x\hat{i} + v_y\hat{j} + v_z\hat{k}$.

- To emphasize the distinction between a vector's components and its magnitude, we write the magnitude of a vector, such as $\vec{F}$, as $|\vec{F}|$. However, when it is obvious that a magnitude is being described, we use the plain symbol (such as F with no coordinate subscript) to denote a vector's magnitude.

- We often choose to spell out the names of objects that are associated with mathematical variables—writing, for example, $\vec{v}_{ball}$ and not $\vec{v}_b$ for the velocity of a ball.

- Numerical subscripts most commonly denote sequential times, positions, velocities, and so on. For example, x_1 is the x-component of the position of some object at time t_1, whereas x_2 is the value of that parameter at some later time t_2. We have avoided using the subscript zero to denote initial values, as in x_0 to denote "the initial position along the x axis," to emphasize that *any* time can be chosen as the initial time for consideration of the subsequent time evolution of a system.

- To avoid confusing the numerical time sequence labels with object labels, we prefer to use capital letters as object labels. For example, we would label two particles A and B rather than 1 and 2. Thus, $\vec{p}_{A1}$ and $\vec{p}_{B1}$ would represent the translational momenta of two particles before a collision whereas $\vec{p}_{A2}$ and $\vec{p}_{B2}$ would be their momenta after a collision.

- To avoid excessively long strings of subscripts, we have made the unconventional choice to write all adjectival labels as *super*scripts. Thus, Newton's Second Law is written $\vec{F}^{net} = m\vec{a}$ whereas the sum of the forces acting on a certain object might be written as $\vec{F}^{net} = \vec{F}^{grav} + \vec{F}^{app}$. To avoid confusion with mathematical exponents, an adjectival label is never a single letter.

- Following a usage common in contemporary physics, the time average of a variable $\vec{v}$ is denoted as $\langle \vec{v} \rangle$ and not as $\vec{v}_{avg}$.

- Physical constants such as e, c, g, G, are all **positive** scalar quantities.

5 Significant Figures and the Precision of Numerical Results

Quoting the result of a calculation or a measurement to the correct number of significant figures is merely a way of telling your reader roughly how precise you believe the result to be. Quoting too many significant figures overstates the precision of your result and quoting too few implies less precision than the result may actually possess. So how many significant figures should you quote when reporting your result.

Determining Significant Figures

Before answering the question of how many significant figures to quote, we need to have a clear method for determining how many significant figures a reported number has. The standard method is quite simple:

> **METHOD FOR COUNTING SIGNIFICANT FIGURES:** Read the number from left to right, and count the first nonzero digit and all the digits (zero or not) to the right of it as significant.

Using this rule, 350 mm, 0.000350 km, and 0.350 m each has *three* significant figures. In fact, each of these numbers merely represents the same distance, expressed in different units. As you can see from this example, the number of *decimal places* that a number has is *not* the same as its number of *significant figures*. The first of these distances has zero decimal places, the second has six decimal places, and the third has three, yet all three of these numbers have three significant figures.

One consequence of this method is especially worth noting. Trailing zeros count as significant figures. For example, 2700 m/s has four significant figures. If you really meant it to have only three significant figures, you would have to write it either as 2.70 km/s (changing the unit) or 2.70×10^3 m/s (using scientific notation.)

A Simple Rule for Reporting Significant Figures in a Calculated Result

Now that you know how to count significant figures, how many should the result of a calculation have? A simple rule that will work in most calculations is:

> **SIGNIFICANT FIGURES IN A CALCULATED RESULT:** The common practice is to quote the result of a calculation to the number of significant figures of the *least* precise number used in the calculation.

Although this simple rule will often either understate or (less frequently) overstate the precision of a result, it still serves as a good rule-of-thumb for everyday numerical work. In introductory physics you will only rarely encounter data that are known to better than two, three, or four significant figures. This simple rule then tells you that you can't go very far wrong if you round off all your final results to three significant figures.

There are two situations in which the simple rule should *not* be applied to a calculation. One is when an exact number is involved in the calculation and another is when a calculation is done in parts so that intermediate results are used.

1. *Using Exact Data* There are some obvious situations in which a number used in a calculation is exact. Numbers based on counting items are exact. For example, if you are told that there are 5 people on an elevator, there are exactly 5 people, not 4.7 or 5.1. Another situation arises when a number is exact by definition. For example, the conversion factor 2.54 cm/inch does *not* have three significant figures because the inch is *defined* to be exactly 2.5400000 . . . cm. *Data that are known exactly should not be included when deciding which of the original data has the fewest significant figures.*

2. *Significant Figures in Intermediate Results* Only the final result at the end of your calculation should be rounded using the simple rule. Intermediate results should never be rounded. Spreadsheet software takes care of this for you, as does your calculator if you store your intermediate results in its memory rather than writing them down and then rekeying them. If you must write down intermediate results, keep a few more significant figures than your final result will have.

Understanding and Refining the Simple Significant Figure Rule

Quoting the result of a calculation or measurement to the correct number of significant figures is a way of indicating its precision. You need to understand what limits the precision of data before you fully understand how to use the simple rule or its exceptions.

Absolute Precision There are two ways of talking about precision. First there is *absolute precision*, which tells you explicitly the smallest scale division of the measurement. It's always quoted in the same units as the measured quantity. For example, saying "I measured the length of the table to the nearest centimeter" states the absolute precision of the measurement. The absolute precision tells you how many *decimal places* the measurement has; it alone does not determine the number of significant figures. Example: if a table is 235 cm long, then 1 cm of absolute precision translates into three significant figures. On the other hand, if a table is for a doll's house and is only 8 cm long, then the same 1 cm of absolute precision has only one significant figure.

Relative Precision Because of this problem with absolute precision, scientists often prefer to describe the precision of data *relative* to the size of the quantity being measured. To use the previous examples, the *relative precision* of the length of the real table in the previous example is 1 cm out of 235 cm. This is usually stated as a ratio (1 part in 235) or as a percentage ($1/235 = 0.004255 \approx 0.4\%$). In the case of the toy table, the same 1 cm of absolute precision yields a relative precision of only 1 part in 8 or $1/8 = 0.125 = 12.5\%$.

Inconsistencies between Significant Figures and Relative Precision There is an inconsistency that goes with using a certain number of significant figures to express relative precision. Quoted to the same number of significant figures, the relative precision of results can be quite different. For example, 13 cm and 94 cm both have two significant figures. Yet the first is specified to only 1 part in 13 or $1/13 \approx 10\%$, whereas the second is known to 1 part in 94 or $1/94 \approx 1\%$. This bias toward greater relative precision for results with larger first significant figures is one weakness of using significant figures to track the precision of calculated results. You can partially address this problem, by including one more significant figure than the simple rule suggests, when the final result of a calculation has a 1 as its first significant figure.

Multiplying and Dividing When multiplying or dividing numbers, the *relative* precision of the result cannot exceed that of the least precise number used. Since the number of significant figures in the result tells us its relative precision, the simple rule is all that you need when you multiply or divide. For example, the area of a strip of paper of measured size is 280 cm by 2.5 cm would be correctly reported, according to the simple rule, as 7.0×10^2 cm^2. This result has only two significant figures since the less precise measurement, 2.5 cm, that went into the calculation had only two significant figures. Reporting this result as 700 cm^2 would not be correct since this result has three significant figures, exceeding the relative precision of the 2.5 cm measurement.

Addition and Subtraction When adding or subtracting, you line up the decimal points before you add or subtract. This means that it's the *absolute* precision of the least precise number that limits the precision of the sum or the difference. This can lead to some exceptions to the simple rule. For example, adding 957 cm and 878 cm yields 1835 cm. Here the result is reliable to an absolute precision of about 1 cm since both of the original distances had this reliability. But the result then has four significant figures whereas each of the original numbers had only three. If, on the other hand, you take the difference between these two distances you get 79 cm. The difference is still reliable to about 1 cm, but that absolute precision now translates into only two significant figures worth of relative precision. So, you should be careful when adding or subtracting, since addition can actually increase the relative precision of your result and, more important, subtraction can reduce it.

Evaluating Functions What about the evaluation of functions? For example, how many significant figures does the sin(88.2°) have? You can use your calculator to answer this question. First use your calculator to note that sin(88.2°) = 0.999506. Now add 1 to the least significant decimal place of the argument of the function and evaluate it again. Here this gives sin(88.3°) = 0.999559. Take the last significant figure in the result to be *the first one from the left that changed* when you repeated the calculation. In this example the first digit that changed was the 0; it became a 5 (the second 5) in the recalculation. So, using the empirical approach gives you five significant figures.

Some Fundamental Constants of Physics*

Constant	Symbol	Computational Value	Best (1998) Value	
			Value[a]	Uncertainty[b]
Speed of light in a vacuum	c	3.00×10^8 m/s	2.997 924 58	exact
Elementary charge	e	1.60×10^{-19} C	1.602 176 462	0.039
Gravitational constant	G	6.67×10^{-11} m³/s²·kg	6.673	1500
Universal gas constant	R	8.31 J/mol·K	8.314 472	1.7
Avogadro constant	N_A	6.02×10^{23} mol⁻¹	6.022 141 99	0.079
Boltzmann constant	k_B	1.38×10^{-23} J/K	1.380 650 3	1.7
Stefan–Boltzmann constant	σ	5.67×10^{-8} W/m²·K⁴	5.670 400	7.0
Molar volume of ideal gas at STP[d]	V_m	2.27×10^{-2} m³/mol	2.271 098 1	1.7
Electric constant (permittivity)	ϵ_0	8.85×10^{-12} C²/N·m²	8.854 187 817 62	exact
Coulomb constant	$k = 1/4\pi\epsilon_0$	8.99×10^9 N·m²/C²	8.987 551 787	5×10^{-10}
Magnetic constant (permeability)	μ_0	1.26×10^{-6} N/A²	1.256 637 061 43	exact
Planck constant	h	6.63×10^{-34} J·s	6.626 068 76	0.078
Electron mass[c]	m_e	9.11×10^{-31} kg	9.109 381 88	0.079
		5.49×10^{-4} u	5.485 799 110	0.0021
Proton mass[c]	m_p	1.67×10^{-27} kg	1.672 621 58	0.079
		1.0073 u	1.007 276 466 88	$1.3 \times .10^{-4}$
Ratio of proton mass to electron mass	m_p/m_e	1840	1836.152 667 5	0.0021
Electron charge-to-mass ratio	e/m_e	1.76×10^{11} C/kg	1.758 820 174	0.040
Neutron mass[c]	m_n	1.68×10^{-27} kg	1.674 927 16	0.079
		1.0087 u	1.008 664 915 78	5.4×10^{-4}
Hydrogen atom mass[c]	m_{1H}	1.0078 u	1.007 825 031 6	0.0005
Deuterium atom mass[c]	m_{2H}	2.0141 u	2.014 101 777 9	0.0005
Helium atom mass[c]	m_{4He}	4.0026 u	4.002 603 2	0.067
Muon mass	m_μ	1.88×10^{-28} kg	1.883 531 09	0.084
Electron magnetic moment	μ_e	9.28×10^{-24} J/T	9.284 763 62	0.040
Proton magnetic moment	μ_p	1.41×10^{-26} J/T	1.410 606 663	0.041
Bohr magneton	μ_B	9.27×10^{-24} J/T	9.274 008 99	0.040
Nuclear magneton	μ_N	5.05×10^{-27} J/T	5.050 783 17	0.040
Bohr radius	r_B	5.29×10^{-11} m	5.291 772 083	0.0037
Rydberg constant	R	1.10×10^7 m⁻¹	1.097 373 156 854 8	7.6×10^{-6}
Electron Compton wavelength	λ_C	2.43×10^{-12} m	2.426 310 215	0.0073

[a]Values given in this column should be given the same unit and power of 10 as the computational value.
[b]Parts per million.
[c]Masses given in u are in unified atomic mass units, where 1 u = 1.660 538 73 × 10⁻²⁷ kg.
[d]STP means standard temperature and pressure: 0°C and 1.0 atm (0.1 MPa).

*The values in this table were selected from the 1998 CODATA recommended values (www.physics.nist.gov).

Some Astronomical Data

Some Distances from Earth

To the Moon*	3.82×10^8 m	To the center of our galaxy	2.2×10^{20} m
To the Sun*	1.50×10^{11} m	To the Andromeda Galaxy	2.1×10^{22} m
To the nearest star (Proxima Centauri)	4.04×10^{16} m	To the edge of the observable universe	$\sim 10^{26}$ m

* Mean distance.

The Sun, Earth, and the Moon

Property	Unit	Sun		Earth	Moon
Mass	kg	1.99×10^{30}		5.98×10^{24}	7.36×10^{22}
Mean radius	m	6.96×10^8		6.37×10^6	1.74×10^6
Mean density	kg/m^3	1410		5520	3340
Free-fall acceleration at the surface	m/s^2	274		9.81	1.67
Escape velocity	km/s	618		11.2	2.38
Period of rotation^a	—	37 d at poles^b	26 d at equator^b	23 h 56 min	27.3 d
Radiation power^c	W	3.90×10^{26}			

^{a} Measured with respect to the distant stars, ^{b} The Sun, a ball of gas, does not rotate as a rigid body; ^{c} Just outside Earth's atmosphere solar energy is received, assuming normal incidence, at the rate of 1340 W/m^2.

Some Properties of the Planets

	Mercury	Venus	Earth	Mars	Jupiter	Saturn	Uranus	Neptune	Pluto
Mean distance from Sun, 10^6 km	57.9	108	150	228	778	1430	2870	4500	5900
Period of revolution, y	0.241	0.615	1.00	1.88	11.9	29.5	84.0	165	248
Period of rotation,^{a} d	58.7	-243^b	0.997	1.03	0.409	0.426	-0.451^b	0.658	6.39
Orbital speed, km/s	47.9	35.0	29.8	24.1	13.1	9.64	6.81	5.43	4.74
Equatorial diameter, km	4880	12 100	12 800	6790	143 000	120 000	51 800	49 500	2300
Mass (Earth = 1)	0.0558	0.815	1.000	0.107	318	95.1	14.5	17.2	0.002
Surface value of g,^{c} m/s^2	3.78	8.60	9.78	3.72	22.9	9.05	7.77	11.0	0.5
Escape velocity,^{c} km/s	4.3	10.3	11.2	5.0	59.5	35.6	21.2	23.6	1.1

^{a} Measured with respect to the distant stars.
^{b} Venus and Uranus rotate opposite their orbital motion.
^{c} Gravitational acceleration measured at the planet's equator.

Conversion Factors

Conversion factors may be read directly from these tables. For example, 1 degree $= 2.778 \times 10^{-3}$ revolutions, so $16.7° = 16.7 \times 2.778 \times 10^{-3}$ rev. The SI units are fully capitalized. Adapted in part from G. Shortley and D. Williams, *Elements of Physics*, 1971, Prentice-Hall, Englewood Cliffs, N.J.

Solid Angle

1 sphere
$= 4\pi$ steradians
$= 12.57$ steradians

Plane Angle

	°	′	″	RADIAN	rev
1 degree =	1	60	3600	1.745×10^{-2}	2.778×10^{-3}
1 minute =	1.667×10^{-2}	1	60	2.909×10^{-4}	4.630×10^{-5}
1 second =	2.778×10^{-4}	1.667×10^{-2}	1	4.848×10^{-6}	7.716×10^{-7}
1 RADIAN =	57.30	3438	2.063×10^{5}	1	0.1592
1 revolution =	360	2.16×10^{4}	1.296×10^{6}	6.283	1

Length

cm	METER	km	in.	ft	mi
1 centimeter = 1	10^{-2}	10^{-5}	0.3937	3.281×10^{-2}	6.214×10^{-6}
1 METER = 100	1	10^{-3}	39.37	3.281	6.214×10^{-4}
1 kilometer = 10^{5}	1000	1	3.937×10^{4}	3281	0.6214
1 inch = 2.540	2.540×10^{-2}	2.540×10^{-5}	1	8.333×10^{-2}	1.578×10^{-5}
1 foot = 30.48	0.3048	3.048×10^{-4}	12	1	1.894×10^{-4}
1 mile = 1.609×10^{5}	1609	1.609	6.336×10^{4}	5280	1

1 angström $= 10^{-10}$ m   1 fermi $= 10^{-15}$ m   1 light-year $= 9.460 \times 10^{12}$ km   1 fathom = 6 ft   1 yard = 3 ft   1 mil $= 10^{-3}$ in.
1 nautical mile = 1852 m   1 parsec $= 3.084 \times 10^{13}$ km   1 Bohr radius $= 5.292 \times 10^{-11}$ m   1 rod = 16.5 ft   1 nm $= 10^{-9}$ m
$= 1.151$ miles = 6076 ft

Area

METER^2	cm^2	ft^2	in.^{2}
1 SQUARE METER = 1	10^{4}	10.76	1550
1 square centimeter = 10^{-4}	1	1.076×10^{-3}	0.1550
1 square foot = 9.290×10^{-2}	929.0	1	144
1 square inch = 6.452×10^{-4}	6.452	6.944×10^{-3}	1

key: 1 square mile $= 2.788 \times 10^{7}$ ft^2 = 640 acres; 1 barn $= 10^{-28}$ m^2; 1 acre = 43 560 ft^2; 1 hectare $= 10^{4}$ m^2 = 2.471 acres.

Volume

METER^3	cm^3	L	ft^3	in.^{3}
1 CUBIC METER = 1	10^{6}	1000	35.31	6.102×10^{4}
1 cubic centimeter = 10^{-6}	1	1.000×10^{-3}	3.531×10^{-5}	6.102×10^{-2}
1 liter = 1.000×10^{-3}	1000	1	3.531×10^{-2}	61.02
1 cubic foot = 2.832×10^{-2}	2.832×10^{4}	28.32	1	1728
1 cubic inch = 1.639×10^{-5}	16.39	1.639×10^{-2}	5.787×10^{-4}	1

key: 1 U.S. fluid gallon = 4 U.S. fluid quarts = 8 U.S. pints = 128 U.S. fluid ounces = 231 in.^{3} 1 British imperial gallon = 277.4 in.^{3} = 1.201 U.S. fluid gallons.

Mass

Quantities in the colored areas are not mass units but are often used as such. When we write, for example, 1 kg "=" 2.205 lb, this means that a kilogram is a *mass* that *weighs* 2.205 pounds at a location where g has the standard value of 9.80665 m/s².

g	KILOGRAM	slug	u	oz	lb	ton
1 gram = 1	0.001	6.852×10^{-5}	6.022×10^{23}	3.527×10^{-2}	2.205×10^{-3}	1.102×10^{-6}
1 KILOGRAM = 1000	1	6.852×10^{-2}	6.022×10^{26}	35.27	2.205	1.102×10^{-3}
1 slug = 1.459×10^{4}	14.59	1	8.786×10^{27}	514.8	32.17	1.609×10^{-2}
1 atomic mass unit = 1.661×10^{-24}	1.661×10^{-27}	1.138×10^{-28}	1	5.857×10^{-26}	3.662×10^{-27}	1.830×10^{-30}
1 ounce = 28.35	2.835×10^{-2}	1.943×10^{-3}	1.718×10^{25}	1	6.250×10^{-2}	3.125×10^{-5}
1 pound = 453.6	0.4536	3.108×10^{-2}	2.732×10^{26}	16	1	0.0005
1 ton = 9.072×10^{5}	907.2	62.16	5.463×10^{29}	3.2×10^{4}	2000	1

1 metric ton = 1000 kg

Time

y	d	h	min	SECOND
1 year = 1	365.25	8.766×10^{3}	5.259×10^{5}	3.156×10^{7}
1 day = 2.738×10^{-3}	1	24	1440	8.640×10^{4}
1 hour = 1.141×10^{-4}	4.167×10^{-2}	1	60	3600
1 minute = 1.901×10^{-6}	6.944×10^{-4}	1.667×10^{-2}	1	60
1 SECOND = 3.169×10^{-8}	1.157×10^{-5}	2.778×10^{-4}	1.667×10^{-2}	1

Speed

ft/s	km/h	METER/SECOND	mi/h	cm/s
1 foot per second = 1	1.097	0.3048	0.6818	30.48
1 kilometer per hour = 0.9113	1	0.2778	0.6214	27.78
1 METER per SECOND = 3.281	3.6	1	2.237	100
1 mile per hour = 1.467	1.609	0.4470	1	44.70
1 centimeter per second = 3.281×10^{-2}	3.6×10^{-2}	0.01	2.237×10^{-2}	1

1 knot = 1 nautical mi/h = 1.688 ft/s   1 mi/min = 88.00 ft/s = 60.00 mi/h

Force

dyne	NEWTON	lb	pdl
1 dyne = 1	10^{-5}	2.248×10^{-6}	7.233×10^{-5}
1 NEWTON = 10^{5}	1	0.2248	7.233
1 pound = 4.448×10^{5}	4.448	1	32.17
1 poundal = 1.383×10^{4}	0.1383	3.108×10^{-2}	1

1 ton = 2000 lb

Pressure

atm	dyne/cm²	inch of water	cm Hg	PASCAL	lb/in.²	lb/ft²
1 atmosphere = 1	1.013×10^6	406.8	76	1.013×10^5	14.70	2116
1 dyne per centimeter² = 9.869×10^{-7}	1	4.015×10^{-4}	7.501×10^{-5}	0.1	1.405×10^{-5}	2.089×10^{-3}
1 inch of water[a] at 4°C = 2.458×10^{-3}	2491	1	0.1868	249.1	3.613×10^{-2}	5.202
1 centimeter of mercury[a] at 0°C = 1.316×10^{-2}	1.333×10^4	5.353	1	1333	0.1934	27.85
1 PASCAL = 9.869×10^{-6}	10	4.015×10^{-3}	7.501×10^{-4}	1	1.450×10^{-4}	2.089×10^{-2}
1 pound per inch² = 6.805×10^{-2}	6.895×10^4	27.68	5.171	6.895×10^3	1	144
1 pound per foot² = 4.725×10^{-4}	478.8	0.1922	3.591×10^{-2}	47.88	6.944×10^{-3}	1

[a] Where the acceleration of gravity has the standard value of 9.80665 m/s².

1 bar = 10^6 dyne/cm² = 0.1 MPa   1 millibar = 10^3 dyne/cm² = 10^2 Pa   1 torr = 1 mm Hg

Energy, Work, Heat

Btu	erg	ft·lb	hp·h	JOULE	cal	kW·h	eV	MeV
1 British thermal unit = 1	1.055×10^{10}	777.9	3.929×10^{-4}	1055	252.0	2.930×10^{-4}	6.585×10^{21}	6.585×10^{15}
1 erg = 9.481×10^{-11}	1	7.376×10^{-8}	3.725×10^{-14}	10^{-7}	2.389×10^{-8}	2.778×10^{-14}	6.242×10^{11}	6.242×10^5
1 foot-pound = 1.285×10^{-3}	1.356×10^7	1	5.051×10^{-7}	1.356	0.3238	3.766×10^{-7}	8.464×10^{18}	8.464×10^{12}
1 horsepower-hour = 2545	2.685×10^{13}	1.980×10^6	1	2.685×10^6	6.413×10^5	0.7457	1.676×10^{25}	1.676×10^{19}
1 JOULE = 9.481×10^{-4}	10^7	0.7376	3.725×10^{-7}	1	0.2389	2.778×10^{-7}	6.242×10^{18}	6.242×10^{12}
1 calorie = 3.969×10^{-3}	4.186×10^7	3.088	1.560×10^{-6}	4.186	1	1.163×10^{-6}	2.613×10^{19}	2.613×10^{13}
1 kilowatt hour = 3413	3.600×10^{13}	2.655×10^6	1.341	3.600×10^6	8.600×10^5	1	2.247×10^{25}	2.247×10^{19}
1 electron-volt = 1.519×10^{-22}	1.602×10^{-12}	1.182×10^{-19}	5.967×10^{-26}	1.602×10^{-19}	3.827×10^{-20}	4.450×10^{-26}	1	10^{-6}
1 million electron-volts = 1.519×10^{-16}	1.602×10^{-6}	1.182×10^{-13}	5.967×10^{-20}	1.602×10^{-13}	3.827×10^{-14}	4.450×10^{-20}	10^{-6}	1

Power

Btu/h	ft·lb/s	hp	cal/s	kW	WATT
1 British thermal unit per hour = 1	0.2161	3.929×10^{-4}	6.998×10^{-2}	2.930×10^{-4}	0.2930
1 foot-pound per second = 4.628	1	1.818×10^{-3}	0.3239	1.356×10^{-3}	1.356
1 horsepower = 2545	550	1	178.1	0.7457	745.7
1 calorie per second = 14.29	3.088	5.615×10^{-3}	1	4.186×10^{-3}	4.186
1 kilowatt = 3413	737.6	1.341	238.9	1	1000
1 WATT = 3.413	0.7376	1.341×10^{-3}	0.2389	0.001	1

Magnetic Field

gauss	TESLA	milligauss
1 gauss = 1	10^{-4}	1000
1 TESLA = 10^4	1	10^7
1 milligauss = 0.001	10^{-7}	1

Magnetic Flux

maxwell	WEBER
1 maxwell = 1	10^{-8}
1 WEBER = 10^8	1

1 tesla = 1 weber/meter²

Mathematical Formulas

Geometry

Circle of radius r: circumference $= 2\pi r$; area $= \pi r^2$.

Sphere of radius r: area $= 4\pi r^2$; volume $= \frac{4}{3}\pi r^3$.

Right circular cylinder of radius r and height h:
area $= 2\pi r^2 + 2\pi rh$; volume $= \pi r^2 h$.

Triangle of base a and altitude h: area $= \frac{1}{2}ah$.

Quadratic Formula

If $ax^2 + bx + c = 0$, then $x = \dfrac{-b \pm \sqrt{b^2 - 4ac}}{2a}$.

Trigonometric Functions of Angle θ

$$\sin\theta = \frac{y}{r} \qquad \cos\theta = \frac{x}{r}$$

$$\tan\theta = \frac{y}{x} \qquad \cot\theta = \frac{x}{y}$$

$$\sec\theta = \frac{r}{x} \qquad \csc\theta = \frac{r}{y}$$

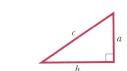

Pythagorean Theorem

In this right triangle,
$$a^2 + b^2 = c^2$$

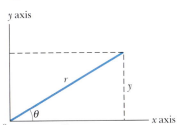

Triangles

Angles are A, B, C

Opposite sides are a, b, c

Angles $A + B + C = 180°$

$$\frac{\sin A}{a} = \frac{\sin B}{b} = \frac{\sin C}{c}$$

$$c^2 = a^2 + b^2 - 2ab\cos C$$

Exterior angle $D = A + C$

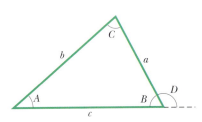

Mathematical Signs and Symbols

$=$ equals

$\approx$ equals approximately

$\sim$ is the order of magnitude of

$\neq$ is not equal to

$\equiv$ is identical to, is defined as

$>$ is greater than ($\gg$ is much greater than)

$<$ is less than ($\ll$ is much less than)

$\geq$ is greater than or equal to (or, is no less than)

$\leq$ is less than or equal to (or, is no more than)

$\pm$ plus or minus

$\propto$ is proportional to

Σ the sum of

$\langle x \rangle$ the average value of x

Trigonometric Identities

$$\sin(90° - \theta) = \cos\theta$$
$$\cos(90° - \theta) = \sin\theta$$
$$\sin\theta/\cos\theta = \tan\theta$$
$$\sin^2\theta + \cos^2\theta = 1$$
$$\sec^2\theta - \tan^2\theta = 1$$
$$\csc^2\theta - \cot^2\theta = 1$$
$$\sin 2\theta = 2\sin\theta\cos\theta$$
$$\cos 2\theta = \cos^2\theta - \sin^2\theta = 2\cos^2\theta - 1 = 1 - 2\sin^2\theta$$
$$\sin(\alpha \pm \beta) = \sin\alpha\cos\beta \pm \cos\alpha\sin\beta$$
$$\cos(\alpha \pm \beta) = \cos\alpha\cos\beta \mp \sin\alpha\sin\beta$$
$$\tan(\alpha \pm \beta) = \frac{\tan\alpha \pm \tan\beta}{1 \mp \tan\alpha\tan\beta}$$

$$\sin\alpha \pm \sin\beta = 2\sin\tfrac{1}{2}(\alpha \pm \beta)\cos\tfrac{1}{2}(\alpha \mp \beta)$$

$$\cos\alpha + \cos\beta = 2\cos\tfrac{1}{2}(\alpha + \beta)\cos\tfrac{1}{2}(\alpha - \beta)$$

$$\cos\alpha - \cos\beta = -2\sin\tfrac{1}{2}(\alpha + \beta)\sin\tfrac{1}{2}(\alpha - \beta)$$

Binomial Theorem

$$(1 + x)^n = 1 + \frac{nx}{1!} + \frac{n(n-1)x^2}{2!} + \cdots \qquad (x^2 < 1)$$

Exponential Expansion

$$e^x = 1 + x + \frac{x^2}{2!} + \frac{x^3}{3!} + \cdots$$

Logarithmic Expansion

$$\ln(1 + x) = x - \tfrac{1}{2}x^2 + \tfrac{1}{3}x^3 - \cdots \qquad (|x| < 1)$$

Trigonometric Expansions (θ in radians)

$$\sin\theta = \theta - \frac{\theta^3}{3!} + \frac{\theta^5}{5!} - \cdots$$

$$\cos\theta = 1 - \frac{\theta^2}{2!} + \frac{\theta^4}{4!} - \cdots$$

$$\tan\theta = \theta + \frac{\theta^3}{3} + \frac{2\theta^5}{15} + \cdots$$

Cramer's Rule

Two simultaneous equations in unknowns x and y,

$$a_1x + b_1y = c_1 \quad \text{and} \quad a_2x + b_2y = c_2,$$

have the solutions

$$x = \frac{\begin{vmatrix} c_1 & b_1 \\ c_2 & b_2 \end{vmatrix}}{\begin{vmatrix} a_1 & b_1 \\ a_2 & b_2 \end{vmatrix}} = \frac{c_1b_2 - c_2b_1}{a_1b_2 - a_2b_1}$$

and

$$y = \frac{\begin{vmatrix} a_1 & c_1 \\ a_2 & c_2 \end{vmatrix}}{\begin{vmatrix} a_1 & b_1 \\ a_2 & b_2 \end{vmatrix}} = \frac{a_1c_2 - a_2c_1}{a_1b_2 - a_2b_1}.$$

Products of Vectors

Let $\hat{i}$, $\hat{j}$, and $\hat{k}$ and be unit vectors in the x, y, and z directions. Then

$$\hat{i} \cdot \hat{i} = \hat{j} \cdot \hat{j} = \hat{k} \cdot \hat{k} = 1, \quad \hat{i} \cdot \hat{j} = \hat{j} \cdot \hat{k} = \hat{k} \cdot \hat{i} = 0,$$

$$\hat{i} \times \hat{i} = \hat{j} \times \hat{j} = \hat{k} \times \hat{k} = 0,$$

$$\hat{i} \times \hat{j} = \hat{k}, \quad \hat{j} \times \hat{k} = \hat{i}, \quad \hat{k} \times \hat{i} = \hat{j}.$$

Any vector $\vec{a}$ with components a_x, a_y, and a_z along the x, y, and z axes can be written as

$$\vec{a} = a_x\hat{i} + a_y\hat{j} + a_z\hat{k}.$$

Let $\vec{a}$, $\vec{b}$, and $\vec{c}$ be arbitrary vectors with magnitudes a, b, and c. Then

$$\vec{a} \times (\vec{b} + \vec{c}) = (\vec{a} \times \vec{b}) + (\vec{a} \times \vec{c})$$

$$(s\vec{a}) \times \vec{b} = \vec{a} \times (s\vec{b}) = s(\vec{a} \times \vec{b}) \quad (s = \text{a scalar}).$$

Let θ be the smaller of the two angles between $\vec{a}$ and $\vec{b}$. Then

$$\vec{a} \cdot \vec{b} = \vec{b} \cdot \vec{a} = a_xb_x + a_yb_y + a_zb_z = ab \cos \theta$$

$$\vec{a} \times \vec{b} = -\vec{b} \times \vec{a} = \begin{vmatrix} \hat{i} & \hat{j} & \hat{k} \\ a_x & a_y & a_z \\ b_x & b_y & b_z \end{vmatrix}$$

$$= \hat{i} \begin{vmatrix} a_y & a_z \\ b_y & b_z \end{vmatrix} - \hat{j} \begin{vmatrix} a_x & a_z \\ b_x & b_z \end{vmatrix} + \hat{k} \begin{vmatrix} a_x & a_y \\ b_x & b_y \end{vmatrix}$$

$$= (a_yb_z - b_ya_z)\hat{i} + (a_zb_x - b_za_x)\hat{j} + (a_xb_y - b_xa_y)\hat{k}$$

$$|\vec{a} \times \vec{b}| = ab \sin \theta$$

$$\vec{a} \cdot (\vec{b} \times \vec{c}) = \vec{b} \cdot (\vec{c} \times \vec{a}) = \vec{c} \cdot (\vec{a} \times \vec{b})$$

$$\vec{a} \times (\vec{b} \times \vec{c}) = (\vec{a} \cdot \vec{c})\vec{b} - (\vec{a} \cdot \vec{b})\vec{c}$$

Derivatives and Integrals

In what follows, the letters u and v stand for any functions of x, and a and m are constants. To each of the indefinite integrals should be added an arbitrary constant of integration. The *Handbook of Chemistry and Physics* (CRC Press Inc.) gives a more extensive tabulation.

Derivatives

1. $\dfrac{dx}{dx} = 1$

2. $\dfrac{d}{dx}(au) = a\dfrac{du}{dx}$

3. $\dfrac{d}{dx}(u + v) = \dfrac{du}{dx} + \dfrac{dv}{dx}$

4. $\dfrac{d}{dx}x^m = mx^{m-1}$

5. $\dfrac{d}{dx}\ln x = \dfrac{1}{x}$

6. $\dfrac{d}{dx}(uv) = u\dfrac{dv}{dx} + v\dfrac{du}{dx}$

7. $\dfrac{d}{dx}e^x = e^x$

8. $\dfrac{d}{dx}\sin x = \cos x$

9. $\dfrac{d}{dx}\cos x = -\sin x$

10. $\dfrac{d}{dx}\tan x = \sec^2 x$

11. $\dfrac{d}{dx}\cot x = -\csc^2 x$

12. $\dfrac{d}{dx}\sec x = \tan x \sec x$

13. $\dfrac{d}{dx}\csc x = -\cot x \csc x$

14. $\dfrac{d}{dx}e^u = e^u\dfrac{du}{dx}$

15. $\dfrac{d}{dx}\sin u = \cos u\dfrac{du}{dx}$

16. $\dfrac{d}{dx}\cos u = -\sin u\dfrac{du}{dx}$

Integrals

1. $\displaystyle\int dx = x$

2. $\displaystyle\int au\, dx = a\int u\, dx$

3. $\displaystyle\int (u + v)\, dx = \int u\, dx + \int v\, dx$

4. $\displaystyle\int x^m dx = \frac{x^{m+1}}{m + 1} \quad (m \neq -1)$

5. $\displaystyle\int \frac{dx}{x} = \ln|x|$

6. $\displaystyle\int u\frac{dv}{dx}\, dx = uv - \int v\frac{du}{dx}\, dx$

7. $\displaystyle\int e^x dx = e^x$

8. $\displaystyle\int \sin x\, dx = -\cos x$

9. $\displaystyle\int \cos x\, dx = \sin x$

10. $\displaystyle\int \tan x\, dx = \ln|\sec x|$

11. $\displaystyle\int \sin^2 x\, dx = \tfrac{1}{2}x - \tfrac{1}{4}\sin 2x$

12. $\displaystyle\int e^{-ax}\, dx = -\frac{1}{a}\, e^{-ax}$

13. $\displaystyle\int xe^{-ax}\, dx = -\frac{1}{a^2}(ax + 1)e^{-ax}$

14. $\displaystyle\int x^2 e^{-ax}\, dx = -\frac{1}{a^3}(a^2x^2 + 2ax + 2)e^{-ax}$

15. $\displaystyle\int_0^\infty x^n e^{-ax}\, dx = \frac{n!}{a^{n+1}}$

16. $\displaystyle\int_0^\infty x^{2n} e^{-ax^2}\, dx = \frac{1 \cdot 3 \cdot 5 \cdots (2n - 1)}{2^{n+1}a^n}\sqrt{\frac{\pi}{a}}$

17. $\displaystyle\int \frac{dx}{\sqrt{x^2 + a^2}} = \ln(x + \sqrt{x^2 + a^2})$

18. $\displaystyle\int \frac{x\, dx}{(x^2 + a^2)^{3/2}} = -\frac{1}{(x^2 + a^2)^{1/2}}$

19. $\displaystyle\int \frac{dx}{(x^2 + a^2)^{3/2}} = \frac{x}{a^2(x^2 + a^2)^{1/2}}$

20. $\displaystyle\int_0^\infty x^{2n+1} e^{-ax^2}\, dx = \frac{n!}{2a^{n+1}} \quad (a > 0)$

21. $\displaystyle\int \frac{x\, dx}{x + d} = x - d\ln(x + d)$

Properties of Common Elements

All physical properties are for a pressure of 1 atm unless otherwise specified.

Element	Symbol	Atomic Number Z	Molar Mass, g/mol	Density, g/cm³ at 20°C	Melting Point, °C	Boiling Point, °C	Specific Heat, J/(g·°C) at 25°C
Aluminum	Al	13	26.9815	2.699	660	2450	0.900
Antimony	Sb	51	121.75	6.691	630.5	1380	0.205
Argon	Ar	18	39.948	1.6626×10^{-3}	−189.4	−185.8	0.523
Arsenic	As	33	74.9216	5.78	817 (28 atm)	613	0.331
Barium	Ba	56	137.34	3.594	729	1640	0.205
Beryllium	Be	4	9.0122	1.848	1287	2770	1.83
Bismuth	Bi	83	208.980	9.747	271.37	1560	0.122
Boron	B	5	10.811	2.34	2030	—	1.11
Bromine	Br	35	79.909	3.12 (liquid)	−7.2	58	0.293
Cadmium	Cd	48	112.40	8.65	321.03	765	0.226
Calcium	Ca	20	40.08	1.55	838	1440	0.624
Carbon	C	6	12.01115	2.26	3727	4830	0.691
Cesium	Cs	55	132.905	1.873	28.40	690	0.243
Chlorine	Cl	17	35.453	3.214×10^{-3} (0°C)	−101	−34.7	0.486
Chromium	Cr	24	51.996	7.19	1857	2665	0.448
Cobalt	Co	27	58.9332	8.85	1495	2900	0.423
Copper	Cu	29	63.54	8.96	1083.40	2595	0.385
Fluorine	F	9	18.9984	1.696×10^{-3} (0°C)	−219.6	−188.2	0.753
Gadolinium	Gd	64	157.25	7.90	1312	2730	0.234
Gallium	Ga	31	69.72	5.907	29.75	2237	0.377
Germanium	Ge	32	72.59	5.323	937.25	2830	0.322
Gold	Au	79	196.967	19.32	1064.43	2970	0.131
Hafnium	Hf	72	178.49	13.31	2227	5400	0.144
Helium	He	2	4.0026	0.1664×10^{-3}	−269.7	−268.9	5.23
Hydrogen	H	1	1.00797	0.08375×10^{-3}	−259.19	−252.7	14.4
Indium	In	49	114.82	7.31	156.634	2000	0.233
Iodine	I	53	126.9044	4.93	113.7	183	0.218
Iridium	Ir	77	192.2	22.5	2447	(5300)	0.130
Iron	Fe	26	55.847	7.874	1536.5	3000	0.447
Krypton	Kr	36	83.80	3.488×10^{-3}	−157.37	−152	0.247
Lanthanum	La	57	138.91	6.189	920	3470	0.195
Lead	Pb	82	207.19	11.35	327.45	1725	0.129
Lithium	Li	3	6.939	0.534	180.55	1300	3.58
Magnesium	Mg	12	24.312	1.738	650	1107	1.03
Manganese	Mn	25	54.9380	7.44	1244	2150	0.481
Mercury	Hg	80	200.59	13.55	−38.87	357	0.138
Molybdenum	Mo	42	95.94	10.22	2617	5560	0.251
Neodymium	Nd	60	144.24	7.007	1016	3180	0.188

Element	Symbol	Atomic Number Z	Molar Mass, g/mol	Density, g/cm³ at 20°C	Melting Point, °C	Boiling Point, °C	Specific Heat, J/(g·°C) at 25°C
Neon	Ne	10	20.183	0.8387×10^{-3}	−248.597	−246.0	1.03
Nickel	Ni	28	58.71	8.902	1453	2730	0.444
Niobium	Nb	41	92.906	8.57	2468	4927	0.264
Nitrogen	N	7	14.0067	1.1649×10^{-3}	−210	−195.8	1.03
Osmium	Os	76	190.2	22.59	3027	5500	0.130
Oxygen	O	8	15.9994	1.3318×10^{-3}	−218.80	−183.0	0.913
Palladium	Pd	46	106.4	12.02	1552	3980	0.243
Phosphorus	P	15	30.9738	1.83	44.25	280	0.741
Platinum	Pt	78	195.09	21.45	1769	4530	0.134
Plutonium	Pu	94	(244)	19.8	640	3235	0.130
Polonium	Po	84	(210)	9.32	254	—	—
Potassium	K	19	39.102	0.862	63.20	760	0.758
Radium	Ra	88	(226)	5.0	700	—	—
Radon	Rn	86	(222)	9.96×10^{-3} (0°C)	(−71)	−61.8	0.092
Rhenium	Re	75	186.2	21.02	3180	5900	0.134
Rubidium	Rb	37	85.47	1.532	39.49	688	0.364
Scandium	Sc	21	44.956	2.99	1539	2730	0.569
Selenium	Se	34	78.96	4.79	221	685	0.318
Silicon	Si	14	28.086	2.33	1412	2680	0.712
Silver	Ag	47	107.870	10.49	960.8	2210	0.234
Sodium	Na	11	22.9898	0.9712	97.85	892	1.23
Strontium	Sr	38	87.62	2.54	768	1380	0.737
Sulfur	S	16	32.064	2.07	119.0	444.6	0.707
Tantalum	Ta	73	180.948	16.6	3014	5425	0.138
Tellurium	Te	52	127.60	6.24	449.5	990	0.201
Thallium	Tl	81	204.37	11.85	304	1457	0.130
Thorium	Th	90	(232)	11.72	1755	(3850)	0.117
Tin	Sn	50	118.69	7.2984	231.868	2270	0.226
Titanium	Ti	22	47.90	4.54	1670	3260	0.523
Tungsten	W	74	183.85	19.3	3380	5930	0.134
Uranium	U	92	(238)	18.95	1132	3818	0.117
Vanadium	V	23	50.942	6.11	1902	3400	0.490
Xenon	Xe	54	131.30	5.495×10^{-3}	−111.79	−108	0.159
Ytterbium	Yb	70	173.04	6.965	824	1530	0.155
Yttrium	Y	39	88.905	4.469	1526	3030	0.297
Zinc	Zn	30	65.37	7.133	419.58	906	0.389
Zirconium	Zr	40	91.22	6.506	1852	3580	0.276

The values in parentheses in the column of molar masses are the mass numbers of the longest-lived isotopes of those elements that are radioactive. Melting points and boiling points in parentheses are uncertain. The data for gases are valid only when these are in their usual molecular state, such as H_2, He, O_2, Ne, etc. The specific heats of the gases are the values at constant pressure. *Primary source*: Adapted fron J. Emsley, *The Elements*, 3rd ed., 1998, Clarendon Press, Oxford (www.webelements.com). Data on newest elements are current.

Periodic Table of the Elements

Legend:
- Metals
- Metalloids
- Nonmetals

IA																	**0**
1 H	**IIA**											**IIIA**	**IVA**	**VA**	**VIA**	**VIIA**	2 He
3 Li	4 Be				Transition metals							5 B	6 C	7 N	8 O	9 F	10 Ne
11 Na	12 Mg	**IIIB**	**IVB**	**VB**	**VIB**	**VIIB**	**VIIIB**			**IB**	**IIB**	13 Al	14 Si	15 P	16 S	17 Cl	18 Ar
19 K	20 Ca	21 Sc	22 Ti	23 V	24 Cr	25 Mn	26 Fe	27 Co	28 Ni	29 Cu	30 Zn	31 Ga	32 Ge	33 As	34 Se	35 Br	36 Kr
37 Rb	38 Sr	39 Y	40 Zr	41 Nb	42 Mo	43 Tc	44 Ru	45 Rh	46 Pd	47 Ag	48 Cd	49 In	50 Sn	51 Sb	52 Te	53 I	54 Xe
55 Cs	56 Ba	57-71 *	72 Hf	73 Ta	74 W	75 Re	76 Os	77 Ir	78 Pt	79 Au	80 Hg	81 Tl	82 Pb	83 Bi	84 Po	85 At	86 Rn
87 Fr	88 Ra	89-103 †	104 Rf	105 Db	106 Sg	107 Bh	108 Hs	109 Mt	110 Ds	111 Uua	112 Uub	113	114 Uuq	115	116	117	118

Alkali metals (IA), Noble gases (0)

THE HORIZONTAL PERIODS (1–7)

Inner transition metals

Lanthanide series *

57 La	58 Ce	59 Pr	60 Nd	61 Pm	62 Sm	63 Eu	64 Gd	65 Tb	66 Dy	67 Ho	68 Er	69 Tm	70 Yb	71 Lu

Actinide series †

89 Ac	90 Th	91 Pa	92 U	93 Np	94 Pu	95 Am	96 Cm	97 Bk	98 Cf	99 Es	100 Fm	101 Md	102 No	103 Lr

The names of elements 104 through 109 (Rutherfordium, Dubnium, Seaborgium, Bohrium, Hassium, and Meitnerium, respectively) were adopted by the International Union of Pure and Applied Chemistry (IUPAC) in 1997. As of May 2003, elements 110, 111, 112, and 114 have been discovered. See www.webelements.com for the latest information and newest elements.

Answers to Reading Exercises and Odd-Numbered Problems

Chapter 34

RE 34-1: (a) Since the induced emf around the dotted loop must oppose the increase in $\vec{B}$, $\vec{E}$ on the right of the rectangle points down in the negative y direction. $\vec{E} + d\vec{E}$ on the left has a greater magnitude and points in the same direction. (b) Since $\vec{E} \times \vec{B}$ must be in the positive x direction, $\vec{B}$ on the right points into the paper in the negative z direction. $\vec{B} + d\vec{B}$ on the left points in the same direction as $\vec{B}$ but has a greater magnitude.

RE 34-2: In the positive x direction.

RE 34-3: For total absorption, $P_r = I/c$ independent of area, but $F_r = P_r A$ so it decreases as the area decreases.

Problems

1. 5.0×10^{-21} H **3.** $B_x = 0$, $B_y = -6.7 \times 10^{-9} \cos[\pi \times 10^{15}(t - x/c)]$, $B_z = 0$ in SI units **5.** 0.10 MJ **7.** 8.88×10^4 m² **9.** (a) 16.7 nT; (b) 33.1 mW/m² **11.** (a) 6.7 nT; (b) 5.3 mW/m²; (c) 6.7 W **13.** (a) 87 mV/m; (b) 0.30 nT; (c) 13 kW **15.** 3.44×10^6 T/s **17.** (a) z axis; (b) 7.5×10^{14} Hz; (c) 1.9 kW/m² **19.** 89 cm **21.** (a) 3.5 μW/m²; (b) 0.078 μW; (c) 1.5×10^{-17} W/m²; (d) 110 nV/m; (e) 0.25 fT **23.** 1.0×10^7 Pa **25.** 5.9×10^{-8} Pa **27.** (a) 100 MHz; (b) 1.0 μT along the z axis; (c) 2.1 m⁻¹, 6.3×10^8 rad/s; (d) 120 W/m²; (e) 8.0×10^{-7} N, 4.0×10^{-7} Pa **31.** 1.9 mm/s **33.** (b) 580 nm **35.** (a) 4.68×10^{11} W; (b) any chance disturbance could move the sphere from being directly above the source, and then the two force vectors would no longer be along the same axis **37.** (a) 1.9 V/m; (b) 1.7×10^{-11} Pa **39.** 3.1% **41.** 4.4 W/m² **43.** 2/3 **45.** (a) 2 sheets; (b) 5 sheets **47.** 0.21 **49.** 35° **51.** 0.031 **53** 19.6° or 70.4° (= 90° − 19.6°) **55.** (a) 0.50 ms; (b) 8.4 min; (c) 2.4 h; (d) 5500 B.C. **57.** (a) 515 nm, 610 nm; (b) 555 nm, 5.41×10^{14} Hz, 1.85×10^{-15} s **59.** it would steadily increase; (b) the summed discrepancies between the apparent time of eclipse and those observed from x; the radius of Earth's orbit

Chapter 35

RE 35-1: a

RE 35-2: $0.2d$, $1.8d$, $2.2d$.

RE 35-3: When you look into a flat mirror, you see the portion of light scattering off your face that bounces off the mirror and travels straight back into your eyes. But you assume that the light entering your eyes has traveled in a straight line to reach you, so you see an image of your face behind the mirror. The image of your face is right side up. The light from your hair hits the mirror at a slight angle and then bounces into your eyes from above which is why you see your hair on top. Left and right are a different story. If you are standing face to face with another person and your right ear points toward the east, her left ear will point toward the east. If, instead, you face a flat mirror, the light from your right ear will bounce off the mirror and enter your eyes from the east. Even though your east ear is the east ear of the image, your right ear has become the left ear of the image.

RE 35-4: Ray 1: A ray that is initially parallel to the central axis reflects as if it came originally from the focal point *behind* the mirror. Ray 2: A ray that comes from the object and is traveling toward the focal point behind the mirror emerges parallel to the central axis. Ray 3: A ray that comes from the object and is traveling toward the center of curvature C of the mirror returns along itself. Ray 4: A ray that comes from the object and reflects from the mirror at its intersection c from the central axis is reflected symmetrically from the central axis.

RE 35-5: (a) Real; (b) inverted; (c) same.

RE 35-6: (a) e; (b) virtual, same.

Problems

1. 1.48 **3.** 1.26 **5.** 1.07 m **11.** 1.22 **13.** (a) 49°; (b) 29° **15.** (a) cover the center of each face with an opaque disk of radius 4.5 mm; (b) about 0.63 **17.** (a) $\sqrt{1 + \sin^2 \theta}$; (b) $\sqrt{2}$; (c) light emerges at the right; (d) no light emerges at the right **19.** 49.0° **21.** 40 cm **23.** (a) 3 **27.** new illumination is 10/9 of the old **29.** 10.5 cm **33.** (a) 2.00; (b) none **37.** $i = -12$ cm **39.** 45 mm, 90 mm **43.** 22 cm **47.** same orientation, virtual, 30 cm to the left of the second lens; $m = 1$ **53.** (a) 13.0 cm; (b) 5.23 cm; (c) -3.25; (d) 3.13; (e) -10.2 **55.** (a) 2.35 cm; (b) decrease **57.** (a) 5.3 cm; (b) 3.0 mm

Chapter 36

RE 36-1: b (least n), c, a.

RE 36-2: (a) 3λ, 3; (b) 2.5λ, 2.5.

Problems

1. (a) 5.09×10^{14} Hz; (b) 388 nm; (c) 1.97×10^8 m/s **3.** 1.56 **5.** 22°, refraction reduces θ **7.** (a) 3.60 μm; (b) intermediate, closer to fully constructive interference **9.** (a) 0.833; (b) intermediate, closer to fully constructive interference **11.** (a) 0.216 rad; (b) 12.4° **13.** 2.25 mm **15.** 648 nm **17.** 16 **19.** 0.072 mm **21.** 6.64 μm **23.** 2.65 mm **25.** $y = 27 \sin(\omega t + 8.5°)$ **27.** (a) 1.17 m, 3.00 m, 7.50 m; (b) no **29.** $I = \frac{1}{9}I_m[1 + 8 \cos^2(\pi d \sin \theta/\lambda)]$, I_m = intensity of central maximum **31.** Fully constructively **33.** 0.117μm, 0.352 μm **35.** 70.0 nm **37.** 120 nm **39.** (a) 552 nm; (b) 442 nm **43.** 140 **45.** 1.89μm **47.** 2.4 μm **49.** $\sqrt{(m + \frac{1}{2})\lambda R}$, for $m = 0, 1, 2, \ldots$ **51.** 1.00 m **53.** $x = (D/2a)(m + \frac{1}{2})\lambda$, for $m = 0, 1, 2, \ldots$ **55.** 588 nm **57.** 1.00030

Chapter 37

RE 37-1: (a) expand, (b) expand

RE 37-2: (a) second side maximum, (b) 2.5

RE 37-3: (a) red, (b) violet

RE 37-4: Diminish

RE 37-5: (a) left, (b) less.

Problems

1. 60.4 μm **3.** (a) $\lambda_a = 2\lambda_b$; (b) coincidences occur when $m_b = 2m_a$ **5.** (a) 70 cm; (b) 1.0 mm **7.** 1.77 mm **11.** (d) 53°, 10°, 5.1° **13.** (b) 0 rad, 4.493 rad, etc.; (c) -0.50, 0.93, etc. **15.** (a) 1.3×10^{-4} rad; (b) 10 km **17.** 50 m **19.** (a) 1.1×10^4 km; (b) 11 km **21.** 27 cm **23.** (a) 0.347°; (b) 0.97° **25.** (a) 8.7×10^{-7} rad; (b) 8.4×10^7 km; (c) 0.025 mm **27.** five **29.** (a) 4; (b) every fourth bright fringe is missing **31.** (a) nine; (b) 0.255 **33.** (a) 3.33 μm; (b) 0.0°, $\pm10.2°$, $\pm20.7°$, $\pm32.0°$, $\pm45.0°$, $\pm62.2°$ **35.** three **37.** (a) 6.0 μm; (b) 1.5 μm; (c) $m = 0, 1, 2, 3,$ 5, 6, 7, 9, **39.** 1100 **47.** 3650 **53.** 0.26 nm **55.** 39.8 pm **59.** (a) $a_0/\sqrt{2}$, $a_0/\sqrt{5}$, $a_0/\sqrt{10}$, $a_0/\sqrt{13}$, $a_0/\sqrt{17}$ **61.** 30.6°, 15.3° (clockwise); 3.08°, 37.8° (counterclockwise)

Chapter 38

RE 38-1: We observe that the second train is moving with respect to our train. The "slight vibration" we feel is evidence that our own train is moving along the tracks, but this does not tell us either the speed or the direction of that motion. Without this information on our own motion, we cannot determine whether or not the second train is at rest with respect to the tracks.

RE 38-2: (a) Our measured value of the speed of light is equal to its value measured by the rider. (b) With respect to our frame, it takes some time for the light to move from one end of the boxcar to the other. During that time the boxcar moves in a direction opposite to that of the light. As a result, we measure the distance between emission and absorption of the light to be smaller than the length of the boxcar. (c) Part (b) shows that the distance between emission and absorption is shorter in our frame than in the frame of the rider on the boxcar. The speed of light is the same for both of us. Therefore, the time between emission and detection is shorter as measured in our frame is shorter than the time measured in the boxcar frame. (You should revisit this analysis after reading Section 38-12 Lorentz Contraction. Will this re-analysis lead to the same conclusion or a different one?)

RE 38-3: These questions concern individual impressions, so there are no objective answers. Here are mine: Halfway through the performance I would experience it as a whole series of events: hard parts, easy parts, mistakes! Those who printed the program probably listed the Minute Waltz as one event in the concert. Looking back ten years later, I will probably (but not necessarily) remember it as a single event.

RE 38-4: (a) Recall that, in general, distance = velocity*time. We know the velocity (c) and the distance (30 meters) of the returning light pulse. Therefore the time taken for this return is (30 m)/(3 $\times$ 10^8 m/s) = 10^{-7} second = 0.1 microsecond. Therefore the pulse arrived at detector B $0.225 - 0.1 = 0.125$ microsecond after it passed us at detector A. (b) The proton pulse left detector A at $t = 0$ and, according to part (a) arrived at detector B at $t = 0.125$ microseconds. Therefore its speed from A to B is (30 m)/(0.125 $\times$ 10^{-6} s) = 2.4 $\times$ 10^8 m/sec, or 2.4/3 = 0.8 of the speed of light.

RE 38-5: Decay reduces the remaining number of pions by a factor of two for every 25 meters of distance they travel (at that particular speed, whatever it is). So there will be one-quarter remaining after 50 meters of travel and one-eighth at a distance of 75 meters from the target.

RE 38-6: All the clocks will run at the rate of every other clock. If this were not so, you could use the difference between rates of different clocks to detect which inertial reference frame you are in, contrary to the principle of relativity.

RE 38-7: Rearrange Eq. 38-3 to read $\Delta\tau/\Delta t = \sqrt{1 - v^2/c^2}$. Square both sides of this equation, solve for v^2/c^2, and substitute the values given in the statement of the exercise, $v^2/c^2 = 1 - \Delta\tau/\Delta t = 1 - 1/1.01$ = 0.0099. Take the square root of both sides to obtain approximately $v/c = 0.1$. This is a rough-and-ready criterion for the speed above which relativistic effects become significant in reasonably accurate experiments.

RE 38-8: The time a light pulse takes to travel one way from Earth's surface to the moon's surface is 3.76×10^8 m/3.00×10^8 m = 1.25 second. The two firecrackers, one on each surface explode one second apart in the earth-moon frame. Nothing, not even light can travel from the first explosion to the second explosion. Therefore one explosion cannot have caused the other one.

RE 38-9: Music has been emitted from the tape player. There are vibrations in the air. This is a fact that must be true in both frames of reference. (For example, it might be arranged to have the noise set off a firecracker, whose explosion must be acknowledged by all.) Air currents and distance permitting, Sam on the ground will be able to hear the music sometime (with what distortions we do not bother to analyze here). When Sam and Susan meet over coffee, they will both verify that some tape has been wound from one spool to the other in the tape recorder.

RE 38-10: Rearrange Eq. 38-17 to read $E/mc^2 = (1 - v^2/c^2)^{-1/2}$. Take the reciprocal of both sides, then square both sides and substitute values for the ratio of energy to rest energy given in the statement of the exercise. The result is $(mc^2/E)^2 = 1/4 = 1 - v^2/c^2$. Rearrange and take a square root to obtain $v = \sqrt{3/4}\,c = 0.866c$.

RE 38-11: The algebraic equations for this solution are essentially identical to those for the solution to the preceding reading exercise 38-10. Rearrange Eq. 38-28 to read $\Delta x'/\Delta x = (1 - v^2/c^2)^{-1/2}$. Take the reciprocal of both sides, then square both sides and substitute values for the ratio of measured lengths given in the statement of the exercise. The result is $(\Delta x'/\Delta x)^2 = 1/4 = 1 - v^2/c^2$. Rearrange and take a square root to obtain $v = \sqrt{3/4}\,c = 0.866c$.

RE 38-12: The light flash will move with speed c in our frame; this is a basic assumption of special relativity (Section 38-3). Verify this result by substituting the values $u' = c$ and $v^{\text{rel}} = 0.9c$ into Eq. 38-31.

$$u = \frac{c + v^{\text{rel}}}{1 + cv^{\text{rel}}/c^2} = \frac{c + 0.9c}{1 + 0.9c^2/c^2} = \frac{1.9c}{1.9} = c \text{ as we predicted.}$$

RE 38-13: Square both sides of Eq. 38-33 and multiply through by the resulting denominator: $(f/f_0)^2(1 + v^{\text{rel}}/c) = (1 - v^{\text{rel}}/c)$. Solve for v^{rel}

$$v^{\text{rel}} = \frac{1 - (f/f_0)^2}{1 + (f/f_0)^2}c = \frac{1 - 0.81}{1 + 0.81}c = 0.1c.$$

Problems

1. (a) $v/c = 3.16 \times 10^{-18}$ (b) $v/c = 9.26 \times 10^{-8}$ (c) $v/c = 2.87 \times 10^{-6}$ (d) $v/c = 10^{-4}$ **3.** EACH of the identical experiments should give the same result in the uniformly moving train as in the closed freight container. **5.** $v/c = 0.990$ or $v = 2.97 \times 10^8$ m/s **7.** You set your clock to the time 2×10^{-4} s. **9.** $\Delta\tau = 4.7 \times 10^{-8}$ s and $\Delta t = 17 \times 10^{-8}$ s. Therefore $\Delta t/\Delta\tau = 3.6$ **11.** (a) 26.3 y (b) 52.3 y (c) 3.71 y **13.** (a) $v/c = 0.995$ (b) 4.8×10^3 m (c) 480 m (d) 48 km (e) 9.8×10^4 particles will survive. **15.** (a) $v/c = 0.9999995$ (b) one year (c) It does not matter as long as the acceleration is small. **17.** (1, 2) timelike, yes; (1, 3) spacelike, no; (2,3) lightlike, yes **21.** 3.51×10^{-8} kg/y or about 35 micrograms/year **23.** 1.4467×10^{-29} kg, or 8.127 MeV **25.** (a) 1.04×10^{10} J (b) 0.116 mg **27.** (a) $v/c = [N(N + 2)]^{1/2}/(N + 1)$ (b) $p = [N(N + 2)]^{1/2} \ mc$ **29.** (a) $m[p^2/(2K)] - [K/(2c^2)]$. For slow particle speed this reduces to the first term, which becomes m, as expected. (b) $m/m_e = 206$ **31.** (a) The lowest total energy after the collision (equal to the total energy before the collision) leaves the products at rest. (b) Kinetic energy of each incident proton is equal to the rest energy (the mass) of one proton. (c) This incident kinetic energy is equal to 1 GeV, which is reasonable since in the zero-total-momentum frame all the incident kinetic energy goes into the creation of mass, provided that the products remain at rest. **33.** $v_x = v^{rel}$ and $v_y = v_y'[1 - (v^{rel})^2/c^2]^{1/2}$ **35.** (a) $\cos\phi = [\cos \ \phi' + v^{rel}/c]/[1 + (v^{rel}/c) \cos \ \phi']$ (b) $\cos \ \phi_o = v^{rel}/c$ (c) $\phi_o = 8.1°$ **37.** (a) $v = 2.6 \times 10^8$ m/s (b) $L = 50$ m. **39.** (a) Yes, at an appropriate speed, proper time between two timelike events can be made as small as desired. (b) $v = 0.999 \ 999 \ 15c$ **41.** velocity with respect to the rocket $= -0.82c$ **43.** Minimum and maximum values occur when daughter particles move along direction of relative motion. $u_+ = 0.990 \ c$ and $u_- = 0.282 \ c$ **45.** $f = 22.9$ MHz **47.** $v^{rel} = 0.96 \ c$ **49.** (a) She does not age at all. (b) Both earth and Zircon age 100 y. (c) 350 y (d) 1200 y on earth **51.** 31.6 s **55.** (a) 0.511 MeV (b) $M_{sys} = m + 2m_e$ (c) Mass of the system is $2m_e$ both before and after the collision. **57.** $E_M = (M^2 + m^2)c^2/(2m)$ **61.** Partial answer: Let T be the time lapse between the instant we see the sun explode and the instant we see Venus change color. Then we have time $T/3$ to escape earth after we see Venus change color. This assumes that the alien ship moves faster than the pulse emitted by the sun.

Photo Credits

Index

Page references followed by italic *table* indicate material in tables.
Page references followed by italic *n* indicate material in footnotes.

Mathematical Formulas*

Quadratic Formula

If $ax^2 + bx + c = 0$, then $x = \dfrac{-b \pm \sqrt{b^2 - 4ac}}{2a}$

Binomial Theorem

$(1 + x)^n = 1 + \dfrac{nx}{1!} + \dfrac{n(n-1)x^2}{2!} + \cdots$ $\qquad (x^2 < 1)$

Products of Vectors

Let θ be the smaller of the two angles between $\vec{a}$ and $\vec{b}$. Then

$$\vec{a} \cdot \vec{b} = \vec{b} \cdot \vec{a} = a_x b_x + a_y b_y + a_z b_z = |\vec{a}||\vec{b}|\cos\theta$$

$$\vec{a} \times \vec{b} = -\vec{b} \times \vec{a} = \begin{vmatrix} \hat{i} & \hat{j} & \hat{k} \\ a_x & a_y & a_z \\ b_x & b_y & b_z \end{vmatrix}$$

$$= \hat{i}\begin{vmatrix} a_y & a_z \\ b_y & b_z \end{vmatrix} - \hat{j}\begin{vmatrix} a_x & a_z \\ b_x & b_z \end{vmatrix} + \hat{k}\begin{vmatrix} a_x & a_y \\ b_x & b_y \end{vmatrix}$$

$$= (a_y b_z - b_y a_z)\hat{i} + (a_z b_x - b_z a_x)\hat{j} + (a_x b_y - b_x a_y)\hat{k}$$

$$|\vec{a} \times \vec{b}| = |\vec{a}||\vec{b}|\sin\theta$$

Trigonometric Identities

$\sin\alpha \pm \sin\beta = 2\sin\frac{1}{2}(\alpha \pm \beta)\cos\frac{1}{2}(\alpha \mp \beta)$

$\cos\alpha + \cos\beta = 2\cos\frac{1}{2}(\alpha + \beta)\cos\frac{1}{2}(\alpha - \beta)$

Derivatives and Integrals

$\dfrac{d}{dx}\sin x = \cos x$ $\qquad$ $\displaystyle\int \sin x\, dx = -\cos x$

$\dfrac{d}{dx}\cos x = -\sin x$ $\qquad$ $\displaystyle\int \cos x\, dx = \sin x$

$\dfrac{d}{dx}e^x = e^x$ $\qquad$ $\displaystyle\int e^x\, dx = e^x$

$\displaystyle\int \dfrac{dx}{\sqrt{x^2 + a^2}} = \ln(x + \sqrt{x^2 + a^2})$

$\displaystyle\int \dfrac{x\, dx}{(x^2 + a^2)^{3/2}} = -\dfrac{1}{(x^2 + a^2)^{1/2}}$

$\displaystyle\int \dfrac{dx}{(x^2 + a^2)^{3/2}} = \dfrac{x}{a^2(x^2 + a^2)^{1/2}}$

Cramer's Rule

Two simultaneous equations in unknowns x and y,

$$a_1 x + b_1 y = c_1 \qquad \text{and} \qquad a_2 x + b_2 y = c_2,$$

have the solutions

$$x = \dfrac{\begin{vmatrix} c_1 & b_1 \\ c_2 & b_2 \end{vmatrix}}{\begin{vmatrix} a_1 & b_1 \\ a_2 & b_2 \end{vmatrix}} = \dfrac{c_1 b_2 - c_2 b_1}{a_1 b_2 - a_2 b_1}$$

and

$$y = \dfrac{\begin{vmatrix} a_1 & c_1 \\ a_2 & c_2 \end{vmatrix}}{\begin{vmatrix} a_1 & b_1 \\ a_2 & b_2 \end{vmatrix}} = \dfrac{a_1 c_2 - a_2 c_1}{a_1 b_2 - a_2 b_1}.$$

* See Appendix E for a more complete list.

The Greek Alphabet

Alpha	A	α	Iota	I	ι	Rho	P	ρ
Beta	B	β	Kappa	K	κ	Sigma	Σ	σ
Gamma	Γ	γ	Lambda	Λ	λ	Tau	T	τ
Delta	Δ	δ	Mu	M	μ	Upsilon	Y	υ
Epsilon	E	ϵ	Nu	N	ν	Phi	Φ	ϕ, φ
Zeta	Z	ζ	Xi	Ξ	ξ	Chi	X	χ
Eta	H	η	Omicron	O	o	Psi	Ψ	ψ
Theta	Θ	θ	Pi	Π	π	Omega	Ω	ω